Ethical AI: Navigating the Future With Responsible Artificial Intelligence

Explore the Ethical Dimensions of AI, Learn How to Navigate Its Challenges, and Embrace Responsible AI Practices for a Better, More Inclusive Future

L. D. Knowings

losses, direct or indirect, that are incurred as a result of the use of the information contained within this document, including, but not limited to, errors, omissions, or inaccuracies.

Table of Contents

Introduction

At the end of 2022 and beginning of 2023, the world was overtaken by the release of what was to become one of the most popular artificial intelligence (AI) tools in the market: ChatGPT. As soon as it was made available to the market, there was a boom of companies interested in the tools and how they could enhance their business positions. As an illustration of the matter, it has been identified that AI represents the most significant spend for almost 50% of top tech executives for companies in the American economy, and the budgets reserved for AI represent more than double the investment made in the second-spending area, cloud technology (Rosenbaum & Anwah, 2023).

This means one certain thing: AI has come to stay. No more is it a thing that pertains only to tech companies, not at all! AI is being integrated into healthcare, customer service, social media management, and almost any business area and industry you can think of. Companies both big and small are investing in enhancing their services with these tools, making it a central element in today's economy. At the same time, the increase of AI tools in the corporate world has prompted fear in employees and ethical questions that most do not seem to know how to answer.

On one side, we have companies that are looking to optimize their operations and become more cost-efficient. Conversely, we have employees claiming that AI is robbing them of their jobs and taking their places in the market. Of course, there are also the extremists who say AI is going to replace humanity and that we will be servants to machines, but we can leave those aside, since this probability is highly unlikely. Finally, we have songwriters, artists, singers, writers, authors, and a wide variety of professionals who claim that AI is using their proprietary information to train machines and mimic their art and content.

This leads us to the question: *How much of this is true? How will AI impact our lives in the near future? What are the things users should be aware of when using AI or implementing them in their practice?* These are all very good questions, though most people are still uncertain of the answers. There is so much information going around from different outlets that it is easy to be confused and even overwhelmed by the task of establishing a foundation for our own opinions. How do we know what is true? Or even more importantly, how do we know that we are using AI ethically and not surpassing the boundaries of what is proprietary information, such as AI images being created based on the style and works of specific artists?

These are the crucial points being discussed today by everyone, and I bet it's one of the reasons why you were attracted to this book. With so much information available, how do we know who to trust? How are we supposed to build a solid understanding of the matter? What is even more important: How can we know that what we are reading is based on facts, studies, and overall reliable information and not on some text generated by an AI machine?

As AI consumers, whether just as casual users or as top managers for companies looking to implement the technology, it is important to understand the limitations, capabilities, and boundaries when it comes to the matter. Not doing so can have various consequences ranging from missing out on an opportunity for your business to being considered outdated in the employment market. This makes understanding how AI works and how it can be used essential, not to say urgent, for keeping up with the ongoing, fast-paced changes.

However, even if implementing AI in your business is pressing, you must remember that maintaining an ethical position is also essential. And this is exactly one of the things you are about to learn in this book. As you read, we will take you on a journey that will help you learn more about AI and understand how to use it for your business to achieve optimal results while maintaining an ethical stance. This book will bring you information that will aid you in implementing the technology while showing you alternatives and applications for different markets.

In addition to this, you will be given a crash course on what AI is, the foundation it is built on, and the different advantages you will gain from starting to use it in your company. We will also talk about regulations (or lack thereof) and what is to be expected of the future. By the end of the book, you will have gained significant knowledge on the topic. You will be knowledgeable enough to discuss the implications and consequences of using AI not only in business but also in life in general.

If you are ready to take a deep dive into the world of AI and discover what it can do for you, wait no longer! Join me in exploring all the opportunities, challenges, and discussions AI has brought to our present and how it is continuously changing our relationship with technology.

Chapter 1:

Getting Into Artificial

Intelligence

Have you ever heard the term artificial intelligence (AI) thrown around and felt completely lost? You're not alone. AI has become a buzzword in this digital era, often confusing many. From those who believe that AI will replace them in their jobs to others who are using the tool to leverage their work, there is still a lot of discussion about what it can do, how to use it, and its benefits and disadvantages.

However, most people are unaware that AI has been a significant presence in our lives already—and for some time now. From the results obtained in a Google search to the suggestions in your streaming service, all of this is AI. Despite not being tagged as AI by the general population, we can say that AI goes as far back as before we had computer software to perform translations, or the embedded functions in word processors to correct grammar while you write a document. Yes, it might be a surprise, but all of this is AI.

While these applications were not initially recognized as AI, it is a known fact today that they were. However, once technology started impacting the tasks and jobs people had, it brought much more attention to its use. To make matters somewhat worse, we have all the movies that show a futuristic "vision" and application of AI based on the minds of movie creators. In many of these, AI is used in robots that will eventually dominate the human race and control everything around us, leaving us doomed.

Many people are scared of AI's repercussions and how it will impact their lives, as well as what to expect in the future, especially as technology evolves. Well, let me be the first to say that these are just movies—fiction. No matter the representation that AI has in movies, it is doubtful (at least in the foreseeable future) that this doomsday scenario will happen. However, it is undeniable that AI will change the way we work and even how we carry out tasks in our personal lives.

Therefore, the first thing that you need is a clear understanding of what AI is, what it does, its history, and its applications. Once this is done, you will have a clearer vision of its potential impacts and how it can be applied in your life. This chapter is your key to unraveling the mystique surrounding artificial intelligence. You will soon clearly understand what AI is and how much of an accessible and exciting field it can be. You will see there is no need to be intimidated at all. In fact, I am sure that by the end of this chapter, you will be anxious to learn more and see how it can be applied—ethically—in several aspects of your life.

What Is AI?

With its myriad applications, you might ask yourself, *Okay, if there are so many applications and uses, what exactly is AI?* In short, AI is a branch of computer science that enables a machine to think and process information as a human would. Let's look at an example: When you use your phone to type a message to someone, your keyboard will, in many cases, show a suggestion of the next word you might want to use in the message, right? This is because the program has "learned" the way you speak and the most common expressions you type, and based on this, it will suggest the next word according to the most frequent or probable sentence structure you have used before.

If you use your phone to fill in personal information on websites, such as your full name, it will "register" the order in which these words are used, and it is likely that if you type in your first name, the suggestion it will bring you is your second name, and so on. This is because the computer has learned how you think and therefore mimics your rationale. Since this intelligence was learned through machine learning techniques (which we will talk about soon), it is therefore called *artificial intelligence*. The exact process can be applied in many applications that we see in our day-to-day.

Think about when you want to carry out a search in the engine of your preference. Sometimes, when you type in the first three words of the search, the browser will bring back suggestions as a list of options that might "predict" what you will ask or search for. This does not come "naturally" to a machine. It has learned this based on the thousands, or maybe even millions, of searches carried out daily. The computer uses a program command, or algorithm, to identify the most probable combinations of these words based on the searches carried out by other users and your navigation history.

This is why, many times, the computer seems to be "reading your mind." No, it's not magic; it is just the machine using logic and the most common uses of the same terms other people have used. This is something that it has learned. It is a similar approach to the suggestions in your favorite streaming service. Why do you "trust" its suggestions so much to help determine the next thing you are going to listen to or watch? Simple—because the computer uses your ratings and analyzes other users' behavior to determine a suggestion you will like. It has *learned* what you like based on previously established actions.

This is nothing more than AI in action. While you might not realize it, AI is *everywhere,* and most of the time, it is harmless, *and* it makes our lives easier. While many people are currently scared of AI and what will happen in the future, you have likely used it at least once without realizing it. Apart from the examples already mentioned, the list is ongoing and not exhaustive. Its applications can be seen in customer chatbots, voice assistants such as Siri or Alexa, filters in social media, banking verification, the ads you see when you enter a website, and so on.

Nevertheless, I can understand if you still have questions or concerns regarding AI. After all, there is a lot of alarm regarding how AI will be used and how it works. Here are some of the most common misconceptions regarding AI you will most likely hear:

- *Computers learn independently without any human help, and they will soon be able to make decisions alone!* Okay, machines do learn with input, I will give you that. However, making this jump from learning based on a dataset to making rationalized decisions as a human would is just not possible, at least not for now. This is because when humans learn and execute tasks, they consider several other factors, such as environment, feelings, and the overall situation. Unless it is possible to place all of these in a machine (which, as of now, it is not), it is unlikely they will be able to create plans to dominate the world. Furthermore, before a machine is "taught" to process information, there is a whole step of human involvement in the process, including selecting the data, filtering it, and much more.

- *Computers never make mistakes and are completely objective!* This is one of the most untrue assumptions regarding AI. AI does make mistakes, and it always has bias. You must consider that it is trained using data selected by humans, and humans make mistakes and are biased. If you prepare an AI with incorrect information, it will bring back wrong information. To name a very current example, we can use ChatGPT. When it was released, it was informed by the platform and the free version was only trained with information available up to September 2021. This means that any additional information after this period was not accounted for. While this is not a mistake *per se*, can you imagine the impact of a person carrying out scientific investigations and prompting the software while missing out on all the most current developments? The same goes for biased information, and so on.

- *AI will replace me in my job!* Despite the initial alarm of the potential harm AI might cause in the workplaces, people now understand that AI will not replace them in certain tasks and might even *help* them with the activities they must carry out. This does not mean that some jobs will not cease to exist. Some obviously will. On the other hand, knowing how to use AI and learning its best uses and applications will open up a whole new field of jobs. For example, doctors might be able to use AI to help identify potential anomalies in images or blood exams that cannot be detected by the human eye, or identify patterns that are not obvious. Does this mean that doctors will not be necessary anymore to analyze these images? Not at all! A computer cannot detect all the nuances that a human can. Furthermore, the machine will need to be taught the different parameters, and a human will eventually need to check the information. This means that the way the doctor works (in this case) might need to be adjusted, but that does not mean we won't need doctors to treat patients or give their input anymore.

Perhaps, despite all you have read so far, you are still not convinced about AI. That is okay and understandable. There are still several doubts and considerations that should be taken into account, which can range from its ethical application to its efficiency. However, allow me to take you on a journey through the history of AI so you can better understand how it has influenced our lives in ways that many people can't even imagine. This might help you understand that AI will be a threat only if we allow it and do not adapt.

The History of AI

It is no secret that since the creation of the first computer and, after that, the internet, technology has developed at an alarming rate. Sometimes, just when we learn about the advances made the week before, an update is released and changes everything. Once again, we are behind. It wouldn't be surprising if this has happened before to you or to anyone you know. With the speed at which new solutions are being developed, we are likely to fall behind in one concept or another.

It seems like not so long ago, we were watching movies in which robots took over the world, computers determined our fate, or everything had to be validated by a machine. This reality was similar to space travel: It seemed so far away, and then, with the blink of an eye, a man was stepping on the moon. The same logic can be applied to sci-fi movies. It seemed unlikely or at least far-fetched that we could have a computer make decisions or resemble the way we think, and yet, here we are in 2023, and so much has changed.

But how did we get to this point? What were the significant impacts in history that brought us to this time and led us to the point at which we are now? These are really good questions; depending on who you ask, the relevant events concerning AI might be the same or completely different.

The words "artificial intelligence" were first mentioned in 1956 by a man named John McCarthy. This was the first time a name was given to the concept, even though several people had said it without assigning a "proper" name. Think about the old Greek philosophers who studied the human mind and how our reasoning was created. The authors wrote about machines that could think for themselves or carry out specific tasks. Granted, for the longest time, all these examples of automation and "intelligent" machines considered by those who came before us were mostly just imagination, something of the "future." Nevertheless, it is undeniable that, to some degree, it referred to the concept of AI as we know it today.

However, it was only in the 20th century, more precisely in the early 1900s, when people started talking about machines with intelligence and referring to artificial people, which we would later refer to as robots. This was the first AI "summer," which refers to increased discussion and application of the concept. When Alan Turing published his work about computer intelligence, *Computer Machinery and Intelligence*, in 1950, this was followed by Arthur Samuel, who created a machine that could play chess. The concept started to advance until it was finally named "artificial intelligence" by John McCarthy. The term gained notoriety and became widely used in the scientific community and shortly after, in books, movies, and other works.

During this period, in the 1950s to the 1960s, there was a lot of discussion and progress in the area, including the development of a program that resembled what we know today as artificial neural networks (ANN), a language to develop code for AI (Lisp), and evolutions of computer programs that were able to understand the human language. There was so much enthusiasm coming from the scientific community on the subject it was believed that by the end of the 1960s, computers would be able to replace humans in the way they think, act, and feel.

Progress was fast, and several applications were developed and implemented throughout various industries. Here, it is important to mention automated machines and the concept of "machine learning," which became popular after being used by Arthur Samuel (the man who created the chess-playing machine). This was a decade when the development of machines, programming languages, software, and applications was at the center of most discussions, bringing to the general public perceptions of what the future could bring, alongside the release of the famous sci-fi movie *2001: A Space Odyssey*, which is still considered a benchmark for sci-fi movies today.

However, much of the "buzz" regarding AI was toned down by the 1970s, and developments were halted, bringing it to the first "winter," or period of lower production in the field. With the lack of development and advances, scientists started experiencing a lack of funding for their projects despite some projects still being maintained. Governments did not want to invest in these technologies and were reluctant to continue research because of potential developments.

The reflection of this "winter" period continued in the early 1980s but was soon remedied, stimulated by projects being carried out by the Japanese and the British. Countries that did not want to be left behind resumed projects, and by the end of the 1980s, the AI community was flourishing again and entering a new summer period. This leads us to the 1990s, when many new products with AI technology were released to the public outside the scientific community, such as the "Furby" toy and a robot that could beat a human in chess. A machine was created to understand human commands during this time, and new AI methodologies were discussed (Rangaiah, 2021).

As the 21st century approached and the fear of the "millennium bug" grew with the arrival of the year 2000, AI entered a fast-paced development phase and extrapolated the limits of the scientific community. Movies on the subject took Hollywood by storm, with several releases in the first decade involving mainstream figures like Steven Spielberg and Will Smith. At the same time, we had the release of intelligent cleaning robots (the Roomba), Google was increasing its investments and participation in the market by enhancing its search engine, and we saw the first investments and research for automated cars.

In 2011, Watson, the IBM-developed robot, beat humans in a Jeopardy challenge to seal the relationship between the general audience and the scientific field. It should also be taken into account that digital products had become more widely available to the public: computers, phones, tablets, and the reach of the internet only made the interest increase. As more people became interested in technology, companies invested their resources in creating digital solutions that could aid in several areas, and universities started expanding their courses to include AI and machine learning in the curriculum.

Without noticing, AI-based solutions started being implemented in different aspects of our lives, making it easier without second-guessing their use or questioning their application. It is safe to say that despite being released in 2020, the release of a "new" ChatGPT at the end of 2022 brought a revolution to the market and AI to the center of many discussions. Companies are now rushing to implement the technology. Professionals are looking to learn as much as possible. Schools are looking for potential applications. Maybe it is still impossible to determine if we are in another AI summer. In a few years, we might know.

In the meantime, we should focus on the present, the now. How are you going to use the available tools in your life? What is the best—and most ethical—application of technology that uses AI? To understand this and make informed decisions, it is important to learn the different types of AI, their uses, and examples of where they are currently applied.

Types of AI

When we talk about AI, we have already seen that it has different applications depending on how it will be used. Usually, establishing the type of AI that will be used will vary according to the capabilities that are needed and what is expected from it. When you have an AI program that will recommend music based on preference, it will not be the same AI "type" that is used in a chatbot. Therefore, the first step toward understanding AI better is to examine the different types of functionalities that can be used.

To better understand the differences between them, here is a comparative table that will help you to better visualize and compare each characteristic:

AI Type	Memory	Capability	Examples
Reactive machines	Only uses present references,	Limited	Deep Blue machine created by

	does not have a memory of past actions.		IBM that beat a human in chess.
Limited memory	Limited and temporary, referring only to certain events in the past.	Limited	Self-driving cars.
Theory of mind*	Extensive. Is able to retain information and react to the environment based on observations.	Limitless	Robots that can mimic human interactions and feelings and recognize emotions to make decisions.
Self-awareness*	Extensive. The computer is able to process feelings on its own, as well as understand what *it is*	Limitless	Robots that are self aware and can control their emotions, characteristics, and even have their own beliefs. The machine will

	feeling and its traits.		not only process information, but it will also have a conscience and be able to establish its own opinions and thoughts.

*Both of these types of AI do not currently exist.

Although there have been developments to create machines that can recognize our feelings and adapt accordingly, we are still distant from using this technology. Two of the main reasons for the delay in creating this technology are the ethical limits and controversies that surround them, such as how these capabilities will be used when machines are able to think for themselves.

Although currently, we only have the first two types of AI mentioned in the previous table, and they have limited capacity, this does not mean at all that they are inefficient. This is because within their scope, they can be used to generate new information or analyze existing information while applying it to real-world situations. When these techniques are applied, it opens the possibilities to what the machine can perform and how it will react to specific situations. To better understand these concepts and their applications, let's take a look at what they do and examples of how they work.

Generative and Applied AI

AI can be used for many purposes, but there are only two different ways to approach the problems it is looking to solve. The developer can either use *generative AI* or *applied AI*, and the choice of which one to use will depend, once again, on the purpose and the business question that needs to be solved. To put each of these different techniques into simple applications, we could say that *generative AI* will create new information based on what it has been trained on, and *applied AI* will work with the data it contains to deliver real-world results based on the analysis made.

Now, this might be rather confusing, so instead of giving you different technical terms (that honestly, the layperson would not understand), let's put this into practical examples and applications. This will make the comprehension and visualization easier to understand and to explain, if needed. We will start with talking about generative AI, what it is, and how it can be applied.

Generative AI

Suppose that you are writing a book and you want to create a cover for it. You have the options of hiring a designer, using an application to create it, or asking an AI application to generate one for you based on the desired parameters. When you prompt the AI program, you will tell it exactly what you need: the type of background, the font, the feeling you want to convey, and the colors you want to be used. Based on all the information it has in its database, the AI program will create a *unique* design that supposedly does not already exist (according to the information it has) and *generate* this new cover.

The same can be applied if you are going to use an AI program to write an article. You will tell it the parameters that you desire, such as tone, points to cover, and style to use, and based on the data it has, it will bring you a unique text. It is, therefore, *generating* information. Some examples that can be mentioned are the most recent programs available in the market, from ChatGPT to image creating software.

Now, this all sounds pretty cool, right? And it is. However, there is one very big issue that is currently at the center of the discussion. Think about this: If the program is creating new information based on existing information and AI programs do not think for themselves, where does this information come from? I will tell you that these machines are trained using the styles, tones, information, and all the other characteristics obtained from the work of *other* people.

Therefore, if you are going to create a cover with a certain style, the design that it will bring you may be unique, but in fact, it is still using different characteristics that belong to other professionals. This means that if you ask the machine to create a drawing that "resembles Picasso," for example, it will use all the characteristics and styles that the Spanish painter used and create a design that mimics almost perfectly their style. While this means that you will have a Picasso-style cover, it also means that you are, in theory, using proprietary information (his style) to create something of your own.

This is where most people find it ethically wrong. While mimicking Picasso will clearly reference the Spanish painter and his characteristics, it will be possible to determine that this is AI because it is "common sense" that he did not create it. But what if we are talking about all the anonymous illustrators, writers, narrators, and other creative professionals who had their content used to train the machine and are now having their work and style copied, mimicked, and reproduced without the appropriate referencing or even credit? I bet you can see where the problem lies.

This is one of the main issues of generative AI. While it is creating unique content, it was trained using information produced by others and, therefore, this means it is not fully original. It would be like, for example, if you uploaded all the books written by your favorite author to an AI program and asked it to create a unique story using the same style they used, but with specific parameters. Now, is this fair? Some say it is because the content is unique and that, in fact, there is no such thing as "original" content. Others say it isn't and that proprietary information is being used without consent.

While the objective of this book is not, as of now, to enter these discussions, I will leave you with some things to think about regarding how this is impacting our lives, but more specifically, the lives of those who are not being fairly compensated for having their work copied. This is a significant discussion currently happening, and a consensus has yet to be reached. Some people have filed lawsuits to prevent their content from being used or to be compensated for the inappropriate use of their proprietary art, but as of now (as of this writing), there has been no ruling from the courts.

Now that you understand the concept of generative AI and the ethical concerns it has raised, it is time to look into the other type of AI, the applied. As you will see, despite being more "limited," it still comes with its own challenges and disputes. But let's take a look at what it is so you can better understand.

Applied AI

As the name might suggest, when you have applied AI, you are going to use the machine's ability to process the information it has been fed to generate analysis and content based on it. While it may seem similar to generative AI, the context of applied AI is simpler because it is not going to generate a new and unique output but rather use the information in the database to bring results that can be applied to the real world.

The advantages of this type of AI include the optimization of processes that previously have needed to be handled by a human and enhanced analysis to find connections between matters that are not immediately evident to humans for several reasons. The use of applied AI is very common in the market and you have likely used it before: Netflix and Spotify recommendations, GPS systems, using Alexa or Siri, or the keyboard in your phone. We could also mention chatbots in customer service, analysis of potential fraud in financial operations, and even analyzing image exams for medical purposes.

Now, you will recognize that these applications are likely to have a limited memory. For example, if you ask your voice assistant what the weather's like today and tomorrow ask what it was that you asked yesterday, it will not remember. On the other hand, if you ask it to create a shopping list for you, it will remember the items you have added until you delete the list (when it will no longer have records). While the machine still "learns" from your behavior, it will not keep the historic information unless it is necessary.

As you can see, its uses are more restricted to the specific issue, parameter, and dataset that was used to train it. This means that if you apply the Google search algorithm to analyze the movements of the stock market, it will very likely not work, since this is not the purpose for which it was developed. The data used is directed, specific, and applied only for certain situations, and therefore, there are limited expectations for its use.

However, *does this mean that applied AI is more acceptable than generative AI?* The answer to this question is: It depends. If you think about all the recent legislation that has been created to protect user data, where the companies need to be *allowed* to use it, it might pose as an ethical matter to some. If you consider the application, for example, in the ads that you receive on social media related to content you previously searched on the internet, this might be disturbing to some.

Lastly, when considering generative and applied AI, you can resort to a very useful and quick reference tool that will help you identify the differences. In generative AI, you will create unique content for which the output will always be different; thus, the reason why the responses in some programs are never the same despite asking the same question. On the other hand, when you use applied AI, the generated input will always bring back the same result, since it is based exclusively on the data and it has no capacity to generate new information.

We have only just begun our AI journey. There is still a lot to learn and examine, so next, we'll delve into the fascinating world of machine learning (ML) and deep learning (DL). Both are fundamental concepts for understanding how the process of teaching and training a machine works in real life. In these cases, you will see that machines aren't just reacting or recalling—they are learning.

Chapter 2:

Foundational Pillars

It might be the case, as you were reading the previous chapter, that the question *How do machines learn?* came to mind. This is a very common question, and if you want to learn about AI, it is a crucial process to understand. While this does not mean that you must understand how algorithms are used or how to program one, it is important to get an overview of its main concepts. After all, the development of AI has machine learning (ML) and deep learning (DL) as its foundational pillars, without which it is almost an impossible task.

Understanding the process that enables a computer to learn is essential to understand why, in some cases, the AI program will show bias and why certain information can be found when other information can't. Nevertheless, the answer to your question of *How do machines learn?* might surprise you. In this chapter, you are going to learn all about how a machine is taught, the different approaches these methods have, and how it is possible for a computer to mimic almost precisely the way humans speak.

Therefore, dive into this chapter to unveil the mechanics of ML and DL. By the time you are done reading, you will have a firm grasp of these foundational AI concepts and their practical applications without having to wrap your mind around technical jargon. You will also read about real-life examples of how each can be applied. This knowledge will enable you to not only better understand the chapters that are to come, but also identify the different situations in which AI is applied.

Understanding Machine Learning

It is not uncommon when listening to people talk about AI that it is used interchangeably with ML. However, this assumption is only partially correct, since ML is one of the components that make up AI but not the only one. This is because when we talk about AI, we are speaking about "a broad term that refers to systems or machines that mimic human intelligence" and that learn based on the data they are provided with (*What Is Machine Learning?*, n.d.). When we talk about ML, the main characteristic that should be taken into consideration is that it uses statistics and specific techniques to ensure that a computer learns what it should do.

If you are thinking this sounds very much like how software works, you are not mistaken. However, it is important to understand that there is a crucial difference between these—and it is not the use of data. According to the Great Learning Team (2023):

> In traditional programming, we would feed the input data and a well-written and tested program into a machine to generate output.

When it comes to machine learning, input data, along with the output, is fed into the machine during the learning phase, and it works out a program for itself. (What Is Machine Learning section)

This means that you are training the machine to be able to not only generate the expected output but also "think" and make decisions based on the patterns it has observed, enabling it to use the same rationale if new data is fed to it. If you consider the incredible processing power computers have to deal with large amounts of data and add to this mix the application of the adequate algorithm, it will be able to determine, based on the data, what happened, what will happen, and what should be done. These three applications are described, respectively, as descriptive, predictive, and prescriptive analysis.

But I guess this still does not answer the question of how a machine effectively "learns," right? So let's get right into it. When speaking about ML techniques, they can be classified into these four different types:

- **Supervised ML:** If you think about the word "supervised," you might immediately think about a process that is being monitored. If you are, then you are halfway there on the road to understanding supervised ML. This is because when this is the learning process that is being applied, there is a human monitoring both what is being given to the machine and the results that are being released. In this case, the person responsible will train the machine with "labeled" (identified) data and the output will be given. However, when the real information is given to evaluate its performance, the individual will give the machine something to analyze that they already know the answer to, so it is possible to check the machine performance.

○ **Example:** We are training the machine to identify fruits and give it the different characteristics of grapes, apples, and bananas. Once the machine is trained and learns to identify each of these fruits, it will then be given a picture of a grape. In this case, you know it is a grape, and this is the expected output from the machine. This means you are supervising the process and evaluating how precise the algorithm is when making its predictions to identify different types of fruits.

○ **Real-life application:** This type of machine learning is best applied when there is an absolute value as the reference. One example is its applications in banks, when the machine is taught to identify fraudulent activities based on the client's previous behavior. This means that if they never spent more than $99.99 on a credit card purchase, if there is one with the amount of $2,000, the machine might identify it as fraud because this does not fit into the pattern for this specific individual.

- **Unsupervised ML:** If supervised learning is a process in which there is the full control of the individual training the machine, when we talk about unsupervised learning, this is exactly the opposite. This means that the dataset is not labeled at all and that the prediction the machine will make is not controlled by a human. In this case, the machine will be given a dataset and, based on what it identifies, will group it according to their patterns, classifications, categories, and similarities, as well as identify the differences between them. This specific approach is used when there is data that does not have any commonality and the machine uses its own criteria (commanded by the algorithm) to establish "connection" points. Contrarily to the application of supervised ML, when you have unsupervised ML, there is no expected answer, and it will depend exclusively on how the machine associates and relates the points in the dataset.

  - ○ **Example:** To make the visualization easier and enable an easy comparison, let's use fruit again. Suppose you have a folder that contains thousands of pictures of exotic fruit you have never seen before. However, you have been asked to sort them, and since you have never seen them, there is no assumption of what is correct and what is incorrect, neither is there a classification criteria. In this case, you are going to use unsupervised ML to sort them according to the criteria the machine establishes as the best. This means that it might group the fruit images by color, size, or even by their shape.

o **Real-life application:** If you subscribe to a streaming service, it is probable that you are using unsupervised learning without even realizing it. Let's suppose that you are a Netflix subscriber. When you opened your account and registered, your "file" in the computer was clean, since you never watched anything. This means that while searching for something to watch, you were given movies and series that belonged to different genres, such as horror, romance, and action.

However, as time passed and you watched and rated what you saw, the machine started to "learn" the types of videos you enjoyed the most and recommend more content that it predicted you might enjoy. This is what the percentage at the top of the page means when it says, for example, 95% for you. It is considering everything you have watched, or the "pattern" you exhibit, to suggest things you will also like. Therefore, if you are a fan of action movies, these are the recommendations that will appear the most instead of horror movies. This is a clear example of how the machine is learning and coming to its own conclusions.

- **Semi-supervised ML:** If you understand the concept of supervised and unsupervised ML, then understanding how semi-supervised ML works and its applications will be easier. We can say that in this case, the dataset is neither fully labeled nor unlabeled; part of it is identified and another part isn't. This means that while the computer will have some input on how it should classify the information, this will only be partial, and it will indeed sort the rest of the data and establish the relationship between them on its own.

  o **Example:** Still referring to the fruit, suppose that within the folder with fruit images you have "common" fruits such as apples and bananas mixed with the exotic "uncommon" fruit. In this case, the semi-supervised ML process would be ideal, since it will have the labeled (known) data like apples and bananas, and it will use this information to separate these images from the others, which will be then classified according to the criteria established by the machine. The person who is supervising will then "label" the output based on the identified characteristics and teach the machine that this is the classification it should consider.

○ **Real-life application:** The application of semi-supervised learning is frequently seen in the healthcare industry. If we are talking about blood exams, for example, the doctors and scientists will select a small sample of diverse results and label them according to what they represent. After this is done, the machine is fed with these results, and it is responsible for identifying anomalies by using the labeled data. Based on this information and on the examples, it will reach conclusions for the remainder of the dataset, proving to be effective in identifying diseases that are just starting to appear in the individual.

● **Reinforcement ML:** This is the fourth and last ML technique, and, just as the others you have seen, the name pretty much defines how it is carried out. In this case, the machine is fed a dataset and will make conclusions according to its criteria. However, in this case, the expected output is known, and if the machine finds the best answer, it will receive a "reward" or a "punishment," both of which will tell it how it performed in a certain process. In this case, they will not have labeled data but instead will learn based on the feedback they receive from the agent, stimulating it to find the optimal results and avoid incorrect assumptions. This is the ML process that might take the longest to be fine-tuned, since it needs extensive practice and feedback before the results can be precise.

- o **Example:** For our last example using fruits, suppose that you have a file with images of exotic and common fruits all mixed together and you want the machine to separate and classify them. However, you do not tell it anything about them and just feed it the data. In this case, if the machine correctly identifies a fruit, it will be given one point. However, if it confuses two different types of fruit, it will lose a point. This is the positive and negative reinforcement put into action.

- **Real-life application:** If you have ever used an online translator and have been asked to rate the quality of the translation, you have actively participated in a reinforcement ML process. This is because when the program asks you to rate the quality of the final text, you are telling it if it did well or not, providing feedback. While it is impossible to know the rewards and punishments that each company applies to the process, they use the rating made by the client to help establish whether the program is doing a good job or not.

As you have seen, ML techniques are applied for uses that we can't even imagine—and this trend is expected to continue growing as companies invest more money to adopt them. According to a recent survey carried out and published in *60 Notable Machine Learning Statistics* (2023), 49% of company executives interviewed claimed they planned to explore or use ML in their business for business analytics (13%), security issues (25%), and sales and marketing (16%). If you add to these executives the 15% of organizations that already use advanced ML techniques, this covers a significant part of the market.

This, however, does not mean it does not come with its challenges. Some of the most significant cited by executives and listed by Korobeyko (2023) include the difficulties in scaling up its potential (43%), reproducing the models accurately (41%), and even getting a company's C-suite to adopt its implementation (34%). Furthermore, there is still the difficulty of creating ML processes that have reduced bias and that are accurate and trustworthy for implementation on large scales. One of the best examples shown by Korobeyko is the 19% rate of accuracy of the face recognition software used by some law enforcement agencies.

Deep Learning Explained

While ML is located under the umbrella of AI and has four different techniques you have just learned about, it is also divided into different subsets. One of these is DL, which is the process of having a machine learn by mimicking how the human brain works. In this case, the task is repeated numerous times with repeated adjustment each time to ensure that there is an optimized result. "We refer to 'deep learning' because the neural networks have various (deep) layers that enable learning. Just about any problem that requires 'thought' to figure out is a problem deep learning can learn to solve" (Marr, 2021).

I can understand this might be confusing, so let's take a step back to better understand the context. Despite DL normally being applied for enormous datasets which are usually not labeled, this does not mean that DL uses only unsupervised ML. Not at all. In fact, depending on the process that needs to be carried out by the machine, it can use any of the four ML techniques you have seen.

The difference here is that while the "traditional" ML application might demand that the individual do some preprocessing and organization of the dataset, this is unnecessary when we are speaking about DL. Additionally, it is important to consider that using this approach to create AI programs requires incredible machine processing power, which may be limited in many companies. For this reason, it is not as common to hear about DL when compared to ML. As you might imagine, this is also a more costly process because of the physical structure it needs, as well as the high level of programming expertise the software's developer needs to have.

The complexity of structuring a DL algorithm is also related to the task you want it to carry out. The main reason for this is because DL will learn according to its experience, just as we do. In this situation, not only does it try to mimic the human learning process but also the physical structure of how our brain works by using a system called artificial neural networks (ANN). When developing a DL algorithm, the ANN will work by producing "attempts to mimic the human brain through a combination of data inputs, weights, and bias. These elements work together to accurately recognize, classify, and describe objects within the data" (IBM, 2023).

The "deep" used in the name is precisely connected to the ANN, since it refers to how many layers make it up. Each "node" represents a human neuron, and it is composed of a minimum of three layers: the input, hidden, and output layers. While all the nodes will have an input and output layer, the number of hidden layers can vary according to the structure and complexity of the program. We could even say that "traditional neural networks only contain 2–3 hidden layers, while deep networks can have as many as 150" (Mathworks, 2019).

Let's look at this approach in a simple way. The information enters the node through the input layer. It then assigns different weights to this information, or in other words, attributes a value to it to give it a certain importance. It is then passed on to the first hidden layer to be analyzed, where it looks into the weight and bias of the information. According to what is identified, it will try to pass the information to the following hidden layer. However, this second layer has a specific determined threshold that needs to be overcome so that the information can "enter" it.

If this happens, it will move on to the next layer that will follow the same process. As this small piece of information goes through each of the layers, they will continue to be measured and weighed, and if the final threshold is surpassed, then this information will go to the output layer. Obviously, the process is more complex than this example, since it involves calculations and different factors that will be used to measure each of the characteristics. Here, the most important thing to remember is that this process enables the output to be more precise as the number of layers increases, since more "analysis" will be done.

Because of all the challenges of creating a DL algorithm and the complexities of developing an ANN, the most common uses we have of this technology come from "big" or wealthy companies, such as Google, Apple, Facebook, and Amazon. The possibility to correctly tag your friends on your Facebook or Google pictures without extensively searching through your list is possible because of DL. Self-driving cars being developed by automotive companies learn the differences between what it should and should not do because of DL algorithms.

Finally, you can see the application of DL in common apps and electronic equipment in voice assistants such as Amazon's Alexa, Apple's Siri, Microsoft's Cortana, and Google's assistant on your phone. They use DL to understand what you are asking and interact with you. However, when we talk about these resources and tools, there is yet another layer of AI that should be taken into consideration, and this is exactly what we are going to see next.

The name is pretty self-explanatory: natural language processing. As we move on to the final section of this chapter, we will explore this aspect of AI together with two other subfields of AI and see how they can be used.

Natural Language Processing and Computer Vision

Understanding the concepts of ML, DL, and ANN is essential if you want to be able to discuss the applications of AI in our lives. *Does this mean that all of the AI we see are based on this?* Well, not all of it, but most of the uses for AI currently use these approaches.

However, this does not mean that they all serve the same purpose, have the same functionalities, or even process information the same way. To understand this, it is important to look into the different ways DL can be applied according to each content. In this case, we are going to look into two significant subsets, or types, of AI: natural language processing (NLP) and computer vision.

The following is what you should know about each of these and an example of how they are applied in our routines.

Natural Language Processing

As you might imagine by the name, NLP is the process of teaching the computer how to understand and respond to the human language. If you are right now thinking about the voice assistants I just mentioned and wondering if they use DL and NLP, you are correct! These assistants, just as chatbots with automatic responses according to what you say, are the perfect example of NLP being used. However, what most people didn't know was that when they were using these, it was in fact an application of AI.

When a computer uses NLP, it is using DL algorithms to learn what words can be used together, the grammatical structure it should use for a specific answer, and even alternatives to correct the sentence structure and content of something you have written. Yes! You read that correctly. When you are using an application to correct a document you created, for example, you are using AI to identify the best way to write your text.

Another example that could be mentioned is how your email is filtered between what needs to go to the spam folder and what needs to be delivered to your inbox. Despite the machine in this case not communicating directly with you, it is using AI to establish whether the message is important or not. In this same context, we could say that when you have software that transcribes an audio file or even others that summarize videos you have seen on the internet, it is using NLP to identify the words used and predict the next ones based on approximation.

Finally, I could not leave out the software that created all the AI hype at the end of 2022 and beginning of 2023: ChatGPT. Can you imagine the amount of data that was used to train it so it could bring accurate results by using a language that is similar to how we speak? Certainly a lot! However, this does not mean that it does not make mistakes or that the information it gives as output is always correct. In fact, even if you consider voice assistants and chatbots, they sometimes misunderstand what we ask (like bringing the wrong music or the incorrect solution to a problem).

This only means that machines still have a long way to go before they are able to perfectly mimic how humans communicate. While we are able to understand and immediately ask what a word that we do not understand means, the computer needs to be fed this information because it cannot simply listen and learn without touching its core database. This means that even if we have achieved great advances in this field, there is certainly a long road ahead before an optimal result is achieved.

Computer Vision

If you are a sports fan, you will easily understand this next AI subset. Have you ever wondered how it is that broadcasting stations are able to determine how many miles a player ran, how many correct passes they gave, and even how many shots were missed throughout a game? Well, the answer is very simple: AI! It is by using a technique called computer vision that these companies are able to measure performance, determine statistics, and see where the tactic can be improved.

In this case, the computer is given the ability of the "human eye," where it can analyze the footing of these events and bring back conclusions, hence the name. In this case, the images will be processed much like sets of numbers or words so that the computer can extract information from them. When applying computer vision, the machine will analyze the pixels of the image for the decision-making process.

Can you think of any other applications for this technology which might be just as popular? If your mind immediately took you to the filters that can be applied to pictures or even the facial recognition feature to tag images, you are correct! These are some other applications of this technology that we commonly use but are not referred to as AI.

While the application of computer vision is not new, it is certainly more advanced than the limited options we were used to seeing before. This is because, according to Mihajlovic (2019), due to the "advances in AI and innovations in deep learning and neural networks, the field has been able to take great leaps in recent years and has been able to surpass humans in some tasks related to detecting and labeling objects."

As we get ready to finish this chapter, can you think of any other applications of AI in our daily lives? Maybe you can even stop momentarily and think about the possibilities *because* of AI. How would your life be different without spell check, for example? Or if you had to sort through hundreds of pages online to find the information you sought? Well, these are just some of the applications most people do not even imagine belong to a category of AI.

However, now that you have the foundational knowledge of ML and DL, you might be wondering, *How else are these technologies used in our daily lives? How are they shaping the industries we know?* From revolutionizing healthcare diagnostics to forecasting financial market trends, AI is no longer a distant dream but a present reality.

Chapter 3:

Beyond the Hype

You have probably reached this chapter feeling amazed by the scope of AI in applications for everyday tools, spanning much further than you could even imagine. If this is the case, you shouldn't worry. It is likely that many people do not know it, or even how to name exactly what the technology being used is—AI. When all the hype regarding the use of AI and its implications surfaced, many could not imagine they were already using it in one form or another. This means you—just like them—might be using AI right now without even knowing it!

Based on the examples you have seen previously, several uses of AI are "hidden" from the general public. From tech to healthcare, AI has infiltrated our lives in more ways than one. This chapter will take you on an "AI journey" as you learn about its different applications in diverse industries. From using Google Maps to help you find the best route to your destination to how the search results are presented on your browser, AI is everywhere—it is just a matter of looking closely to see where.

AI in Technology

When you think about AI, the first thought that comes to mind is likely something related to computers, machines, technology, and even sci-fi movies such as *The Matrix*. Let me be the first to say you are not wrong, especially if you consider that AI is generated from computer processing and ML. However, AI is also applied in the industry to help those dealing with complex processes and to streamline production.

For example, did you know that before software is released to the public, it undergoes a rigorous testing process to see if any issues or errors might affect the user experience? Well, it does, and this process was previously carried out by teams of testers who worked to ascertain the quality of the final product and its usability. With the development of AI tools, certain aspects of this testing process have become faster and more precise, eliminating the possibility of human error or overlooked issues.

These tools also help software developers be more efficient when coding their programs. This means that with a prompt (command) to an AI application that has been trained with coding data, the developer can ask it to write code based on the determined parameters. By doing this, writing the code becomes faster, and the occurrence of mistakes is smaller. In addition to this, even if the developer writes the code themselves, they can submit it to the AI program to evaluate its quality and search for mistakes and opportunities to optimize it. If you think about it, depending on the AI tool we are talking about, even you and I, without any coding experience, could write code to develop a game or a program of some type.

AI can also help companies enhance their safety protocols, creating solutions to problems related to cybersecurity and attacks. As you have seen, this is already used when your email provider filters your messages between what is spam and what is not, but it can also be used to identify potential threats contained in malicious messages. If we were to expand this rationale, we could even mention using AI to identify and eliminate "bots" that interfere with overall internet applications, ranging from those that spread "fake news" to others that overload a website's traffic.

All the examples you have seen are just a small sample of how AI can be applied within the technology industry. It is likely that as time passes and the techniques evolve, this technology will become intrinsically related to the software development process and all that it refers to. From the delivery of more efficient programs to apps that can mimic human behavior with incredible precision, AI will certainly revolutionize how we create, interact with, and produce technology.

AI in Finance

According to Velazquez (2023), it is estimated that the finance industry has already invested over $9.4 billion in AI technology as of 2023 and is forecast to increase this number by 16.5% by the end of 2030. For a clearer picture of the matter, 54% of financial institutions with over 5,000 employees claim to use AI within their processes (University of San Diego, 2021). This has enabled them to save money by automating several tasks and aiding with processes such as fraud detection and establishing creditworthiness among their clients.

The applications of AI in this industry have been present for a long time. If you have ever had to go through an "investment profile questionnaire" to identify the best investment options for you, that was AI in the works. Based on your answers to the questions, the AI determined your investor profile and recommended the best options according to your preferences. In this case, the machine algorithm used the different investment possibilities and characteristics to match you to a tailor-made recommendation.

Another situation in which AI is being used, and you might not even have thought of it, is establishing your loan eligibility. Based on the input of your personal financial information, such as your financial history and credit score, the machine predicts whether you are a potential client for a loan or not. It considers all your characteristics and helps the institution define a personalized option for you, including the value and period over which the loan needs to be repaid. When you are simulating a loan possibility on the bank app or receiving notification of an automatically approved loan, this is the work of AI.

In addition to this, banks use AI to detect and prevent fraud, determine the amount of overdraft you will be allowed to have, and even aid you in separating your expenses into categories. However, financial institutions are not only made up of banks but also of hedge fund companies, brokerage firms, and credit companies. As you might imagine, all these also use AI to improve their performance.

In brokerage firms, AI is used to identify and predict trends for what will happen in the market and the best time to invest. Hedge funds use this to determine what investment will bring the most profit and to aid in diversifying their portfolios. It also enables clients to have customer service 24 hours a day by using chatbots to help solve problems and streamline processes, such as account opening and documentation analysis.

If you recognize any of these services, you will see that AI has been present in the financial industry for a long time—it just wasn't named as such. As technology evolves and new AI programs are developed, it is almost certain that the financial industry will be part of the group that helps bring innovative solutions to the market. However, as we move on to the next industry, you will also see how AI has been used for a long time in medicine, helping doctors and scientists develop solutions that can save lives.

AI in Healthcare

A long time ago, when the internet was just becoming popular, a new program that allowed users to describe their symptoms and receive a diagnosis became a buzz. This was around the end of the 1990s, and it had already demonstrated that AI was in full use in the healthcare industry to aid patients in identifying what they were experiencing. However, only a part of this technology was released to the public, though it had already been in use for decades. Healthcare has long been a consumer of AI to help run diagnostics, identify potential health issues, and determine the best course of treatment for a disease.

A long time has passed, and AI capabilities have increased, enabling robots to analyze image exams and help doctors identify the onset of diseases that cannot be determined by the human eye. It is also applied in the pharmaceutical industry to help establish the effectiveness of a drug in fighting off a certain disease, understand the reliability of vaccines, and even develop tailor-made solutions to expedite patient treatment.

In addition to these applications, AI has helped doctors and nurses streamline patient documentation and organize information. Based on the application of algorithms to an individual's file, it is possible to reduce the time dedicated to administrative tasks and increase the time dedicated to healthcare. This has a significant impact on the quality of treatment people receive and their effectiveness.

However, AI in medicine is still controversial in some aspects. Similarly to the program mentioned in the beginning of the section, many individuals are using AI to replace medical evaluation and come up with self-diagnoses. In many cases, this is an extremely dangerous practice, since computers still cannot replace a human and understand all the nuances involved in identifying a disease, especially those that are life-threatening. This means that despite AI having significant importance for the healthcare community, it is still vital to seek out human specialists for medical issues and precise identification of what is going on.

AI in Manufacturing

Look around you for a minute and observe the things in your environment. It is likely that more than 90% of what you see was manufactured by a machine. These machines were created and programmed to ensure that you get products with the best quality while helping the company save money. This is AI at work.

Manufacturing industries use AI to help them organize and find the best way to use the raw materials to avoid product waste, increase product quality, and optimize production. By using AI to map the manufacturing process, for example, it is possible to determine the need to produce more or less items because of the identified market trends based on historical data. Business owners can run diagnostics on machines and determine if maintenance is needed or if adjustments are necessary to optimize cash flow.

Essentially speaking, when AI is applied to the manufacturing industry, it helps companies obtain the best results by analyzing, mapping, and programming their activities. By running simulations in AI programs, it is possible to forecast potential problems that might occur during the manufacturing process or with the product's quality without even turning on the machine.

When these solutions are applied in the industry, it provides a more cost-effective process which can produce a decrease in prices alongside advantages for the final clients.

AI in Retail and Ecommerce

When the COVID-19 pandemic struck at the end of 2020 and beginning of 2021, the world saw a revolution in the way shopping was done. Without the option of leaving their homes, millions of people turned to the internet to purchase what they needed, from their monthly grocery lists to clothes, electronics, and furniture. Regardless of whether these companies had planned to incorporate AI into their businesses then or at a later date, the pandemic created an urgency for this to happen right away.

Despite larger companies, such as Amazon, already having a recommendation system that guides customers toward products associated with what is in their carts, other online businesses saw the need to incorporate similar techniques immediately to mitigate the risk of losing their positions in the market. Innovations included such things as creating digital replicas of environments. These tools allowed customers to visualize furniture in their rooms or see how a certain piece of clothing would look on them before choosing to make a purchase.

In addition to this, because people were spending more time on the internet, these companies saw the need to find ways to attract more potential customers to their websites for more purchases. In comes targeted ads and personalized marketing to create personalized solutions that seem "perfect" and "just what you need." If a product was left in your shopping cart, a notification was sent to you reminding you that it was still there waiting for you. Email notifications with similar products and suggestions based on previous purchases were also sent. And these marketing strategies show no sign of going away any time soon.

Businesses also adopted AI for implementing dynamic pricing strategies and to optimize their logistic processes. Inventory management was improved and delivery routes planned to reduce fuel consumption. The implementation of virtual personal shoppers helped clients 24 hours a day to ensure a purchase was made. The organization of products presented on websites was optimized based on client preferences, and other tactics to attract buyers were implemented, all using AI.

Among all the industries that are adopting AI, it is safe to say that ecommerce and retailers were part of the group that had to make fast changes to remain relevant in the market. Even if today we are back to a version of "normal," these modifications remain, and those who implemented them were not left behind. This is yet another industry that adopted AI technology (albeit in the rush) and was forever transformed.

AI in Education

Together with diagnosing infirmities, the use of AI in education has generated an outpouring of ethical discussions. When ChatGPT was first released, schools from all over the globe saw a significant increase in school work being produced by students using AI. This generated discussions about its use and the immediate need for educational institutions to find solutions and tools to deal with it. In came the "AI detector" software and other programs that had as their primary objective the identification of work that had been produced by computers.

Now that some measures have been put into practice, educators are looking to AI to help them improve the student experience and make their administrative work more efficient. AI technology and NLP is being used to help teachers identify potential plagiarism, grade homework, and even create personalized tutoring programs for those who need it. In addition to this, these programs enable teachers to give helpful feedback on the quality of what is being produced without expending a significant amount of their time, allowing them to focus on teaching.

In addition to this, teachers are using AI to help them plan lessons, identify different learning opportunities, and develop strategies to enhance the learning process. This includes finding new material to support classes, creating games that can be played to help ingrain the content, and bringing more audio and visual material to the classroom. Overall, it is likely that the learning experience will change significantly with the application of these tools at all educational levels, from primary school all the way to universities and specialized learning programs.

At the administrative level, schools are using AI to compile data on student and staff performance and even in admission processes. Some universities are using recommendation systems to guide students toward subjects that best fit their profiles and interests, as well as the subjects that might enhance their learning journey. Other activities such as budgeting, scheduling, and facilities management are being optimized with the use of AI, granting administrators more time to focus on students and education.

AI in Agriculture

Agriculture is one of the pillars that helped evolve society to what it is today, marking the shift when our ancestors changed from hunting and foraging to establishing settlements and growing what they needed to consume. As time passed, techniques to improve and optimize production were developed, enabling farmers to manage farms on a larger scale. Therefore, it should not be a surprise to learn that AI is also being used in this area to ensure greater yields and improve product quality.

From soil analysis and preparation to the harvesting phase, AI can be seen in all production steps. AI analysis can identify the products and quantities that should be applied for optimal growth, the best season to plant due to the weather condition analysis of historical data, and even the crop forecast based on real-time information. These are just some of the applications of the technology in an industry that will require an increase of 70% by 2050 to fulfill the populational demand globally (Jain, 2021). Thus, the need to be as productive as possible.

When you consider the agricultural environment, it is possible to identify several other applications that go beyond the cultivation process. Companies are using AI to determine the best products to apply to kill/prevent diseases and their correct amounts. Robots and machines are programmed with technology that reduces waste and sorts the product during harvest. Drones are being used to monitor fields and identify any potential problems.

When looking into agriculture from other perspectives apart from crops, we see AI permeating these areas as well. Poultry growers are using technology to help determine the correct heating of the hen houses and to identify uncommon behaviors exhibited by the animals. Livestock producers use AI to monitor animal health and maintenance, such as controlling the daily amount of food and water consumed. By studying these metrics, these farmers are able to identify potential issues with the cattle, for example, and prevent the spread of contagious diseases that can quickly spread among the animals.

AI in Security

Although we have already seen in the first section of this chapter the different applications of AI in cybersecurity to help companies mitigate threats, the technology also has a broader application in ensuring physical safety. One of these uses includes the installation of "smart" cameras that carry out surveillance and aid with facial recognition in densely populated areas, such as train stations and airports. Along with identifying individuals, these cameras are also equipped to detect suspicious activities, items, and behaviors that may be perceived as threatening.

If you think about the potential destruction that a contaminating agent can cause if maliciously released, the use of AI becomes a crucial tool in ensuring that the correct measures are taken in time. These programs will quickly analyze and identify infectious diseases and biological agents, allowing for fast reaction and crisis management to prevent loss of life. In addition to this, AI can help in the decision-making process, forecasting the potential spread of diseases and deciding on precautionary actions to contain their spread.

Other applications of AI in security include technology used in homes and for children monitoring, adoption in the military to amplify defense or attack strategies, police surveillance, and crime rate monitoring in cities. These actions can vary, ranging from data analysis to estimating event occurrence to using technology in identifying targets of an operation. By studying patterns and identifying trends, law enforcement can use AI as a tool to prevent crime and reduce significant threats.

Lastly, AI can also help in matters that are not directly related to human action, such as oil and gas leaks and malfunctions in nuclear energy plants. By adopting these systems for early identification, accidents can be prevented and human lives preserved by forecasting catastrophic events and stopping them before they occur. These might be some of the most important applications of AI despite still having a long path to follow. There is still the need to fine-tune the software and remove bias from the identification systems to ensure an impartial analysis that does not target a specific group.

AI in Entertainment

You already know that AI is used in the recommendation systems used by streaming services for the movies and songs that are shown to you. However, did you know that AI is also used in several other areas of the entertainment industry? If your first thought is the use of special effects and image-generating software, you are correct! This is one of the many uses of AI by professionals in the industry.

Now, if you have ever watched a video on YouTube and asked for the captions to be shown so you can read what is being said, that is also AI use! In this case, the machine generates the content based on speech recognition according to the closest word it identifies.

This is the reason why, in many cases, there are more incorrect words when compared to a human transcription; since a computer carries out the process, it does not have the necessary knowledge or understanding to determine what the person meant to say, only to interpret the sound.

Another application of the technology in the industry has to do with identifying copyright infringement. By using speech recognition software, AI can recognize where there is unauthorized use of an individual's voice, text, or even image. Furthermore, AI can help develop scripts and storylines, build character traits, and even write music when given a specific topic and feeling you want to convey.

In the entertainment industry, there has been significant discussion about the use of AI to replace professionals in several areas: from creating AI-generated characters to star in movies to writing books and songs, many professionals have protested its unlimited use. This even includes a debate over using books and other proprietary content to train the machine without previous authorization, but we will learn more about its ethical implications in Chapter 7.

For now, as you have seen, the potential applications of AI are vast and varied, spanning numerous industries, and are actively reshaping how we live and work. But how do we turn these potential applications into reality? How can these sophisticated AI models be designed and implemented? As we move on to the next chapter, you will learn how to create an AI program from scratch and what considerations to include based on its purpose. Shall we take a look?

Chapter 4:

From Theory to Practice

Whether you are a business owner, school administrator, scientist, or any other professional, you might be interested in adopting AI into your routine to help you complete one or more tasks, optimize a process, or evaluate performance. These are all applications that AI can be used for. As you know, even if you are an artist who wants to train a machine to help you with your initial sketches of your drawings, it can be done with AI. However, when you want to implement any new process or procedure into your activities, you should first thoroughly learn the steps involved to improve your chances of success.

You don't need to be a developer or a computer expert to follow the process. You do, however, need to understand how it works so you can guide the company you hire to do this for you or instruct the individual inside your organization on what it is that you want to accomplish. If you are a decision-maker, it will be essential to learn the needed elements for the project to work—not only to manage your and the stakeholders' expectations but also to evaluate the final results.

Therefore, in this chapter, you are going to "get your hands dirty" and be guided step-by-step through the process to design and implement an AI solution. You will learn the basic concepts of creating an AI program, from data preparation to model deployment. By the end of this chapter, you will have gained practical skills to design and implement AI solutions, as well as awareness of common pitfalls and best practices. Are you ready to put theory into practice?

Steps to Implement AI Solutions

While you are not going to effectively learn how to code an AI program with this book alone, since it is neither practical nor its objective, you will be given a "crash course" in understanding each of the phases involved in the process. To do this, I have prepared a list of the steps with a description of their importance and application. These are usually followed in the order in which they will be presented, although during the process, the developer may go back to a certain step and start the process over.

All these steps are independent but interconnected, and each one will directly impact and influence the subsequent action. Here are the steps that need to be followed for the creation and launch of an AI program:

1. **Determine the problem you want to solve.** This is the first step that should be taken, which is to identify where AI can be beneficial or what problem it can solve. The application will depend on the industry and purpose, meaning that the reasons can significantly vary according

to each necessity. Some examples might include the creation of a chatbot to respond to your clients automatically, a facial recognition program for security purposes, a tool that helps optimize the creation of social media posts for your clients, or even a speech recognition solution that will give the GPS different voice options. As you can see, the possibilities are endless. Determining the problem or business matter for which AI will be adopted also needs to be done according to the available data to train the machine. There is, for example, no use in wanting to use different voice tones for the GPS machine if you do not have samples to train it with.

2. **Gather the relevant data.** After determining the business area or the problem that AI will solve, it is time to gather the relevant data. In this case, the data can be sourced from what is stored in your databases (first-party data), data that you obtain or buy from shared sources (second-party data), and data that comes from unknown sources or data banks (third-party data). It is important to know that when dealing with data, the first-party type is the most secure and reliable set, and these parameters decrease as you move down a step. In addition to this, you must ensure that you have enough data to divide it into training and testing sets, which are

two steps we will discuss soon, and that it is diverse and comprehensive, since AI requires millions of data points to work efficiently. It is possible that after the data is cleaned, processed, and tested, you will need to gather more data to achieve your objective. In this case, you will come back to this step and carry out the full process again.

3. **Prepare the data for processing.** Once you have collected all the data you think you will need, it is time to prepare the data. This part of the process is called "data cleaning" and will involve removing blank spaces, errors, and any other irregularity found within the datasets that might lead the machine to make incorrect predictions. By ensuring that the data cleaning process is correctly carried out, the ML operation will produce the best and most reliable results.

4. **Select the model that will be used.** The model used will be directly related to the purpose of the AI program you want to implement. In this case, as you have seen, the options can be supervised, unsupervised, semi-supervised, or reinforcement learning. The decision and selection will be based on the available data, the expected output, and the time and resources available, to name a few

conditions. Once this has been established, it is time to get the model working.

5. **Train the model.** The first thing the developer will do is separate the data between training and testing sets and use the first group to train the machine. By using computer algorithms in the selected programming language, they will write a series of commands, determining what the machine needs to do.

6. **Evaluate the model.** After the model has been trained, the other group of data, the testing data, will be used to evaluate the ML process. This usually means preliminary results will be shown using graphs, statistics, and even a demonstration of what the AI program can do. Based on the provided output, you might decide that some adjustments must be made, more data added, or even that the desired approach is incorrect (in the worst-case scenario).

7. **Fine-tune the parameters and specifications.** It is likely that after the model has been evaluated, it will need to be fine-tuned to meet expectations and produce the desired outcome. This is the last step of the process before the model is deployed and put into production. In this phase, the AI program is almost ready to be used and implemented in the business.

8. **Deploy the AI model.** When the model is considered finished, it will be deployed, or "put to work." It is in this phase that it is installed and implemented in the business and used by the relevant stakeholders involved in the process.

9. **Make predictions.** Now that the model is implemented and working, it will be used for making predictions and for any other purpose it was designed for. It is possible that during this process, some adjustments might be needed. However, these are likely minor, and the program should be running without any issues.

10. **Update and maintain the model.** Though the machine is constantly learning while it is being used, it will still need to be updated. If you create a new product for your business, for example, the machine will need to be trained with this new data to ensure that the chatbot can present it as an alternative to the client asking questions. Because most businesses evolve and change, this final step of the process may need to be done more or less frequently according to the need. In this case, you will go back to step 2 and start the process again by adding the new data.

These 10 steps you have just read are a *must* during the creation of AI software. Regardless of whether you are going to develop the solution within the organization or outsource it to a specialized company dedicated to this service, the process will be the same. If the option is to use a third-party vendor, knowing these steps will help you follow the process and understand what was done and what is missing.

On the other hand, if you are going to develop the AI solution in-house, there are several other considerations and precautions to be taken to ensure you are not wasting time, money, and other company resources. Considering the elements you are about to read will also help you avoid common pitfalls and mistakes usually related to AI implementation. When this happens, the results can be disastrous. Let's look at what these are:

- **Understand the business need:** When you decide to implement AI software in your business, you must be sure that this is an adequate solution for the organization. Sometimes, there are other possibilities that are less costly and more effective, and will not demand as much time. It is crucial to evaluate whether the business model also "supports" an AI solution based on the market it is involved in, the client profile, and even the other programs that are used. We could say that if the area in which the AI software will be applied does not have updated systems that can support its use, updating these should be your primary focus.

- **Identify risks and challenges:** AI does not come without its challenges and risks. When writing a plan for creating and implementing this solution, you should also list the potential difficulties that might be faced throughout the process. One of the most common risks is related to cybersecurity and information management. While we will talk more about this subject in the next chapter, it is among the risks that should be considered. Other risks include project failure, lack of engagement, and other events that must be taken into account, no matter how big or small they are.

- **Ensure there is enough data and that it is relevant:** Similarly to the GPS example that was previously given, it is important to carry out a preliminary analysis to determine if the data available is adequate to be used in your AI program. If you want to create a recommendation system based on preferences and there is no product rating on your site, this means that you do not have the specific data related to your products. While this information might be obtained from competitors or other companies, they will not be as precise or as relevant as if you used data obtained from your public. In this case, the first step would be to adjust your systems to incorporate this feature and only then think about AI.

- **Evaluate the IT structure:** An AI program is software, and this means it will need an adequate structure to be stored, run, and managed. This means that you should evaluate your IT structure and check if there are any changes that must be made. This can include verifying and possibly upgrading cloud storage capacity, purchasing powerful servers or other hardware on which the program can be run, and whether there is compatibility between existing systems and the AI project, to name a few. Once again, this assessment needs to be made before engaging with the new project since if any alerts are identified, you must solve them first to ensure the adoption of the program is technically feasible.

- **Consider the software's final user:** Many companies want to implement an AI solution in their businesses, but they usually do not consider those who are going to use it. When this happens, the costly program becomes obsolete simply because no one is using it, since it does not match their profile. Whether we are talking about internal personnel or clients, carrying out a study that identifies the target user's needs, preferences, and habits will be essential for the decision-making process.

- **Calculate savings vs. investment:** Sometimes, all the previously mentioned elements are in order, and there is no impediment to starting your project. However, there is an additional consideration that must be made, which is how much will be saved by implementing the solution versus how much it will cost. This means that if the AI program will bring savings of 10% per year, but its cost will surpass this percentage, then it is likely not worthwhile. The study of savings vs. investment needs to be carried out, considering all the elements important for the equation. These may include hiring specialized personnel, hardware investments, maintenance and data costs and comparing them to an increase in the number of clients or purchases, a decrease in personnel, streamlining or increasing production, cost reduction with raw material, and others.

- **Invest in hiring a team of experienced professionals:** While the number of professionals dedicated to creating AI programs is increasing, those specialized in the area are still scarce. In addition to this, due to the lack of professionals with enough expertise, their salaries tend to be exponentially larger than that of a "regular" software developer. Even if you are thinking about hiring a seasoned software developer with no ML experience, you will need to invest in costly training material and courses to ensure that your objective is achieved. To make the decision even harder, you will have no certainty that once this professional is trained, they will continue with your company, as they may receive a better work offer with another company.

- **Establish KPIs, plans, milestones, and monitoring metrics:** Many professionals who decide to implement AI programs suffer from one of two issues: They either micromanage the process and want updates with an unnecessary frequency, or they leave the full management of the project to the IT department or professional who will develop it. Neither of these cases is adequate for AI development programs. Constant updates might halt the process that, as you have seen, is lengthy and requires some time and dedication. A lack of updates can lead to a loss of control of the development, and certain adjustments might be missed. In this case, it is crucial that you establish metrics, indicators, milestones, and a plan that must be followed. These should be made together with the development team or validated by them, since they will be the ones directly involved in the project. Updates should be timely, such as weekly or biweekly, to follow what is being done without impacting the process.

- **Expect periodic maintenance and updates:**
As mentioned earlier, your AI software will, at
one moment or the other, need to undergo
maintenance procedures and updates to its
content. This means that there might be
additional costs involved or even an impact on
the way the business works. If this is the case,
you will need to develop a plan and a budget to
deal with these events and even an emergency
fund to support unforeseen situations. This
value can be established based on a percentage
of the total cost of the project.

- **Start small:** Implementing an AI solution to
our businesses is exciting! It is a significant
change, and it shows competitors, clients, and
the market that *we are surfing the right wave.*
However, it is important that you take
precautions to hold back the excitement and
avoid risking a major disaster. This means that
even if you have a "big" project, you should
start small with a pilot program or even
consider a small application to see how it will
work. It is possible that after the solution is
developed, you will end up identifying that it
was not needed or that it is not adequate for
your business. Therefore, starting with a "test"
of sorts will help you measure its efficiency, use,
and application, and help you decide whether

you'll need to consolidate or adjust for implementation within the business.

If you believe that you have taken all the precautions and followed the steps and tips provided herein, then it is possible that you are ready to implement an AI solution in your business. However, before we move on to talk about AI and cybersecurity, I want to take a moment and explain in more detail some of the steps we have seen before. In the following sections of this chapter, we are going to take a closer look into four crucial steps of the ML process, which, if inappropriately applied, can lead to project failure and investment loss. Therefore, read on to learn more about data wrangling, feature engineering, model training, and deployment.

Data Wrangling and Feature Engineering

An AI program, as you already know, is based on data. Therefore, it is crucial to the development process that the content that will be used to train the machine is treated and cleaned before it is used. In addition to this, once the data has been fed to the machine, it is critical that the parameters used to analyze it are correctly adjusted for optimal results. The input of incorrect or unprocessed data and the inability to correctly tune the machine's parameters can result in errors and lack of precision in the program.

Data Wrangling

Data wrangling is the process of cleaning, organizing, and treating the raw data before it is processed. When you are dealing with information directly from the source, it is unlikely that it is ready to be processed and correctly understood by the machine. Due to this, it is essential that the person who is going to create the AI program treats it and organizes it. For this process, there is no defined set of rules that should be carried out, since each professional has their own method for approaching it.

Essentially speaking, the data wrangling process is carried out after the data has been collected and is composed of different steps that include, according to Stobierski (2021),

- **studying** the data to see what kind of information can be obtained and identifying the best way to process it.

- **structuring** or organizing the data so that it is understandable by the machine and each parameter can be correctly identified. This might mean adding names to the columns, dividing information, and carrying out any treatment needed so it can be used. In this case, the data is going to be put in the format and shape that is needed for the analysis to be made.

- **cleaning** the data to remove any errors, blank spaces, potential duplicates, alien characters, and any other item that might interfere with the correct processing.

- **enriching** the data to determine if there is any additional information that can be added to bring value to the dataset. This process is *optional* and will be needed according to what the developer deems could improve the dataset for processing.

- **verifying** that the information is correct, consistent, and has the adequate quality for processing. In this case, developers will usually ask the machine for a "preview" of the information to ensure that it is as expected and that it can be correctly read. This process can also be carried out with a visual inspection of the data in the spreadsheet or table where it is located.

After all these steps have been carried out, it is possible to say that the data is ready to be used for publication or analysis, which is when it will be given to the machine. There are many benefits for this process, including the possibility to better understand the content of what you are dealing with and ensuring the consistency of what is being used. In addition to this, if you consider that the data will be organized for processing, it will help save time and money for the organization.

However, it is important to understand that the process is lengthy and it can take some time, depending on the size of the dataset. If you have different sets coming from distinct sources, for example, this means that you will need to standardize them all into one format that fits the business demands. This part of the process is usually the one that demands the most from the developer, since they will need to understand what the data represents and how it will be used.

While the tools used by the developer to carry out this process might differ, there is lots of different software available that can help them to do it. Some might prefer using programs especially developed for this purpose that are available online, while others just resort to the mundane Excel spreadsheet. This part of the process is intertwined with feature engineering, which is the next topic we will explore and which is crucial for the correct performance of the AI program.

Feature Engineering

It is possible to say that the feature engineering process happens simultaneously with the data wrangling process, despite being a separate procedure. This step will happen during the data preprocessing, in which there will be the "creation, transformation, extraction, and selection of features, also known as variables, that are most conducive to creating an accurate ML algorithm" (*Feature Engineering*, n.d.). This means that you will not only be treating the data but looking into possibilities where the data can be transformed so the ML process is optimized and the model enhanced.

This means that you will, for example, standardize the currency used in the database to match the desired output. By analyzing the different data available, you can also identify potential new variables that might be useful for the ML process, such as subtracting a value from another and creating a new variable with the name "price variation." This analysis and identification can be made if the developer has expertise in the domain they are working with by visualizing the data in graphs or charts, and even by applying statistical calculations to the preliminary data (Patel, 2021).

Some of the techniques mentioned by Patel (2021) that can be used in the process include

- new categories or numbers that will handle missing values within the dataset or creating a new relationship between them. This process is also known as *imputation*.

- treating outliers in the data that might affect processing. This includes replacing values, capping values at a maximum or minimum, removing the information, or converting these data points into discrete data that will not influence the final result.

- scaling the data so that it is enough to be used and preventing overfitting or underfitting the model.

- standardizing the information so that all the data is within the same type. Some examples include transforming all weights to the same measurement unit, amounts to the same currency, the use of commas and periods, and other information that might need to be converted.

Essentially speaking, you will be adding artificial features or assigning non-native characteristics to the dataset based on observations and identification of needs that will improve the algorithm's performance. While this is a simple step to carry out and demands only basic observation and analysis skills, it is often overlooked because the developer is focused on transforming the data instead of obtaining conclusions from it. When feature engineering is correctly applied, "the resulting dataset is optimal and contains all of the important factors that affect the business problem. As a result of these datasets, the most accurate predictive models and the most useful insights are produced" (Patel, 2021).

Despite treating the data being essential for the ML process, we should also not forget that the model training and deployment of the software are also key elements of this process. Therefore, it is crucial to understand how this is done and the needed steps so that once complete, the software can be deployed for use.

Model Training and Deployment

When you have the data ready, it is time to feed them to the machine so it can analyze, interpret, and draw conclusions from it. However, this process is done in steps: First, the machine is trained and then tested. This is done by dividing the dataset into the training and the testing set in a proportion established by the developer. The training set is usually smaller than the testing set, and these can have varied ratios such as 30/70 or 20/80, depending on the size of the dataset.

It is during the training of the model that the algorithm parameters will be adjusted, and it will be identified if the machine is learning according to the expected. It is also at this moment that the developer will determine the ML model that will be used; supervised, unsupervised, semi-supervised, or reinforcement. The machine training will be essential in establishing how the data will be processed and the quality of the output.

Once the training is done, it will then be time to test the model and see if the analysis is being correctly made. This is when the developer will use the other part of the dataset to evaluate if the machine is generating the adequate output. The quality, efficiency, and precision will be studied and compared to other results depending on the ML learning chosen method. The program is then fine-tuned if needed and, once it is assessed and identified that the optimal output is being reached, it is time for deployment.

Deployment, as you might imagine by its name, is the process of putting the software into production and using it for the designed purpose. If you are creating a chatbot, for example, it will be the phase in which it is implemented on the company website to interact with clients. However, as you might imagine, this is an extremely complex process, since it involves many business departments, ranging from the IT team of developers to the customer service agents who will monitor its performance.

It is essential that during the deployment, all the involved personnel are effectively participating in the process, which can include the integration with other company software, training employees, and elaborating on the relevant documentation, all while identifying whether the AI program is meeting the objectives for which it was created. In the beginning of this process, there will need to be a close observation of how it is working and the quality of the inputs. The main reason for this is that, despite all the testing that is carried out, it is almost impossible to predict what the user behavior will be and the types of prompts that will be given.

In this case, the monitoring will need to be carried out according to the complexity of the program and its reach. If you are dealing with a limited software that only carries out specific tasks and analysis, then this evaluation might be shorter and more spaced. On the other hand, if you are deploying a program as "data-heavy" as ChatGPT, for example, it is likely that the model will need to be continuously monitored by the business agents and software development.

Finally, it is worthwhile mentioning that in some AI programs, the user will be able to provide feedback regarding the quality and accuracy of what is being predicted. This feedback can be used in both ML adopting the reinforcement method, in which the model will automatically receive feedback from the user, or this information can be stored into a database for future developer analysis. Regardless of how this process is being carried out, it is almost certain you have experienced it before: evaluating the quality of a translation in a software, the accuracy of the grammar and spell check in a document, or even to confirm if the recommendation was correctly given to you.

You have now learned the foundational concepts and seen the real-world implications of AI. However, as with any powerful technology, AI has its challenges and risks. One area of particular concern and intrigue is the intersection of AI and cybersecurity. The double-edged sword of AI is a tool that enhances security measures but also poses potential threats that can exploit our growing reliance on digital infrastructure. How can these issues be addressed? What precautions should be taken? To address these questions and others that might be on your mind, let's move on to the next chapter and explore this intrinsic relationship and understand what to expect.

Chapter 5:

AI and Cybersecurity

Cybersecurity has long been a concern for both companies and individuals with the increase in the number of hacker attacks and malicious bots, and even the spread of "fake news." Unfortunately, where AI is concerned, it has been both a potential threat and a critical tool for cybersecurity. At the same time as it can be used to create deep fakes, place people in impossible situations, and create numerous types of problems, it can be used to identify potential safety issues by using facial recognition and threat identification.

It is exactly this fascinating paradox that this chapter will explore: how AI is being used both as a tool for good and for bad, and some of the measures that are being used to prevent misuse. Furthermore, we will also take a brief peak into what we can expect of AI concerning cybersecurity in the future and its potential uses. By the end of this chapter, you will understand and be able to discuss this intersection of AI and cybersecurity, as well as the concept of AI as a Service (AIaaS).

AI as a Cybersecurity Threat

When you look at a picture or video online, how can you be 100% certain that what you are seeing is what really happened? The answer is simple: You can't. As AI has been developing and evolving, it has become harder by the day to separate between real and artificially created images. With a simple search on your browser for AI-powered image editors, you will have around 700 *billion* results, giving you just a small glimpse of what this market has become.

While some people use this for fun, such as changing bodies to look a certain way in a picture or even using filters to enhance one feature or the other, others use it for malicious purposes. Imagine, for example, what would happen if an altered image of a politician doing something inappropriate was used during a campaign, and they could not prove it was a deep fake? We don't even need to go that far. What if someone edited a photo or video of you or a loved one in the same situation? Politicians, to use the previous example, oftentimes have the means to fight this, but what if we are talking about a layperson?

This is only one of the ways in which AI is currently being used as a weapon. It is also worth mentioning, for example, hackers programming elaborate software to attack other systems by exploiting its vulnerabilities. The main reason for this happening is simple: Implementing AI software to defend systems from these attacks is expensive. You must think about the cost of having a specialized professional who will develop and implement the system, its updates, and maintenance, and establish the security features needed.

In this case, we are talking not only about very specific professionals who are specialists in machine learning and cybersecurity but also a professional that has experience dealing with these threats. Even then, "if an AI model is trained to respond to these threats, the data used for such training must be representative of the entire population to the degree to which each threat made up the population" (Osakwe, 2023). The field is so specific (and broad at the same time) that companies and people who work in this area are usually highly paid, making it an unaffordable option for smaller businesses, for example.

Other threats that should be mentioned when considering AI are discussed by Coatesworth (2023), who claims that the data used in these programs should be properly evaluated to avoid bias and misuse of data that might lead to corporate espionage. In addition to this, they emphasize that with the complexity and the value of these programs increasing, especially when developed by large companies, they become a target to be stolen and used for malicious purposes by only making certain adjustments. While we will address the matter of bias in the next chapter, looking into these matters and what is being done to prevent these are of extreme importance.

Consider that when using NLP, for example, messages and links containing threats can be created by simulating a friend talking to you, and it is possible you will not notice. It is likely that the number of cyberattacks will increase if the technology is not applied as a security tool—not even the traditional methods will suffice. AI has the power to adjust and learn and, due to this, it becomes harder to fight.

This means that the only tool that is able to fight AI is AI itself. "Businesses can use AI to enhance the security threat hunting process by feeding it with high volumes of application data. AI software can evaluate threats and implement the most effective strategy to fight against them" (Fraudwatch, 2021). This means using AI to identify the vulnerabilities in the company's systems, which at the moment might only be available to some. Adding other measures such as transparency to the data selection and training processes, as well as the implementation of tools that will help companies identify these attacks, is also beneficial.

However, not everything is bad news. As mentioned earlier, AI can and is being used not only to prevent threats but also as a tool to make physical and online spaces safer for people. Due to its incredible ability to process large amounts of data, its decision-making capability, and its adaptability, it can also be used for good, and this is what we are about to see.

AI as a Cybersecurity Tool

The importance of AI as a cybersecurity tool is so relevant that according to a report mentioned by Sajid (2023), "the market size for Artificial Intelligence in cybersecurity stood at 7.58 billion dollars in 2022 and is expected to reach 80.83 billion by 2030." This represents an increase in 30.1% to fight off the evolution of malware being developed by hackers. Sajid states that approximately 94% of the malicious programs that were identified in 2019 were almost impossible to detect, since they could modify their source or, in other words, learn and adapt to the security systems they were facing.

Just as the process of introducing new data and agents to malicious AI software can generate more powerful programs, the same can be said for cybersecurity tools. If the data used to train the machine is comprehensive and can enhance its decision-making process of identifying between the different variations of the values (true positive, false positive, false negative, true negative), this will represent a significant impact in the industry.

Some of the examples where AI is being used in cybersecurity include, for example, analyzing the behavior of a program or user and enabling the software to identify potential abnormalities that would indicate a threat. These learning processes can also teach the machine how malware operates and to act during an attack by analyzing its patterns and predicting how it will behave. There is also the possibility of using an AI program that is trained with malware to simulate attacks on a company's systems and identify potential vulnerabilities that could be explored.

By using these tools, it is possible to improve items such as network and firewall security and suggest solutions to be applied based on what is identified. Let's take an example of security logs that register potential threats that a system might suffer. According to Moisset

(2023), "traditional security log analysis relies on rule-based systems that are limited in their ability to identify new and emerging threats." On the other hand, if AI is applied, this means that it will be able to analyze and compare different patterns that could indicate a threat, especially if they are items that don't have an immediate relationship between them. This will give the company more insight into potential threats that weren't even identified and that can be prevented or mitigated.

Finally, the last aspect of applying AI as a cybersecurity tool that should be considered is the cost. Despite having mentioned in the previous section that this is a costly process, it is also one that, over time, will help the business *decrease* their costs with cybersecurity and malware protection services. Think about it this way: If hackers are constantly finding new ways to breach systems and ensure they get valuable data, what will happen if a company does not upgrade their systems?

If you consider large organizations such as Google, Amazon, banks and investment companies, retail companies, research facilities, and even hospitals, how valuable is their information to their business? What would happen if they lost the data they have compiled throughout the years? It is possible to say that this amount is almost incalculable. Therefore, applying AI to their security systems might mean a considerable investment, but if they do it correctly and maintain it, it will enable them to save money in the future and decrease the potential costly effects that an attack may have.

AI as a Service (AIaaS)

Have you ever heard of SaaS, or software as a service? If you haven't, this is a tool where a company pays to use a software and its infrastructure for a fraction of the price that it would cost to develop one. In other words, it is software that has a purpose and is not developed for a specific company, but rather can be applied for several uses. This is a common use in the market, and considering AI, you shouldn't be surprised that a new market is emerging: the one of AIaaS, or AI as a service.

This means that companies can hire for cybersecurity reasons, for example, software that has been created by a third party and implement it in their company. This is a positive aspect of its implementation, since most of the code used for malware, phishing, and other threats are usually the same and only have minor variations depending on the business. "For companies that can't or are unwilling to build their own clouds and build, test, and utilize their own artificial intelligence systems, AIaaS is the solution" (Kidd & Miller, 2023).

In a practical application, a company who works with translations could use an AIaaS to help with their processes. Or another could use a specific program to implement NLP and chatbots in their customer service. These are all applications of AIaaS, since these are not developed by them but are still used within their structure. While this means there is a significant reduction in costs with development, hiring, implementation, and maintenance, there are also risks that must be considered.

The first that should be mentioned are the potential threats to cybersecurity, since you will need to share your data with the service provider so it can be adjusted to your needs. Still on this same note, your company will be highly dependent on their services, which means that if troubleshooting is needed, it might take some time, since they also will have other clients. Finally, it is important to consider that when you are giving this data to a third-party vendor, the possibility of having a transparent process is unlikely. This means that since you will not have control over how it is managed, it is impossible to understand the processes and tools used to train the ML program.

Future of AI and Cybersecurity

After you have read all the information regarding AI and its applications in cybersecurity, you might be asking yourself, *What does this mean for the future of AI and its use to help cybersecurity?* Well, while this still remains to be seen as investments in the AI field continue to grow, it is already possible to see some applications in the market.

Today, to name a few, we have AI to search, map, and identify vulnerabilities in systems, potential threat analysis and malware detection, and even things as simple as telling you how strong or weak your password is for a certain platform. Looking into the future, we could talk about broadening the AI scope to match the efforts made by hackers and malicious programs. It is almost certain that companies will continue to invest more in systems that are efficient in identifying abnormalities that could present a threat to their structure and operation.

It is also possible to say that in an expansion of AI application, we could mention these software taking action to defend the systems instead of just identifying and alerting. With the potential of ML and its capabilities, it is possible to imagine that there will be a time in which the machines will be able to take action to protect the systems they were designed for. Maybe that might sound far-fetched, but it still remains to be seen. Nevertheless, this high level of cybersecurity by using AI systems will depend on the core of what makes it so efficient: data.

As we've explored the role of AI in the cybersecurity landscape, it's crucial to remember that these powerful systems aren't infallible. There is a certain amount of bias that is inevitably introduced in AI systems, either from the data they learn from or the way they are programmed. How can we understand and mitigate these biases? What are the ethical implications? Read on to the next chapter to learn more about AI bias and how to minimize the issues they might bring in certain applications.

Chapter 6:

Understanding and

Mitigating AI Bias

Can machines be biased? The answer is not as straightforward as you might think. While humans have biases, whether they are aware of them or not, this determination in machines has several nuances that have been at the center of discussion regarding AI. "At a time when many companies are looking to deploy AI systems across their operations, being acutely aware of those risks and working to reduce them is an urgent priority" (Manyika et al., 2019).

In this chapter, you will learn about the concept of bias in AI, its implications, and strategies to mitigate it. You will see how AI can be unconsciously embedded in an AI algorithm, affecting its development and performance in certain cases. These can be a reflection of the individual who gathered the data or even what the applied criteria is. For these reasons, among others, it is essential to understand how bias can be seen in AI and the actions that can be taken to mitigate it. By the end of this chapter, you will be equipped with knowledge and strategies to address AI bias, one of the significant ethical concerns associated with AI.

What Is AI Bias?

Think about the following: When data is being gathered for a ML program that will generate AI software, who is responsible for carrying out this process? As you have seen in this book, it is usually made by a *person* who determines the data that should be gathered based on the business problem or solution that will be applied. Based on this, and the fact that humans are naturally biased, isn't it possible to say that the gathered data will have bias to some degree?

Despite all the efforts that might be made to decrease its occurrence, it is likely that, in at least some way, the data will have an "opinion," so to speak. And if machines learn based on this data, it is likely to be biased as well. This bias can include things like the way the NLP is processed, such as reflecting the specific way a certain group speaks, how AI image recognition programs have an idea of what is considered "optimal," or even how certain keywords that are considered by analysts as "essential" to establish certain characteristics are applied.

This means, in short, that most of the bias that comes from AI will be generated from the data it is fed and the feedback it is given from users. In addition to this, other uses such as the choice of using data based on historical disparities and prejudiced connections can lead the data to make incorrect predictions (Larkin, 2022). Other "more common" issues, such as the lack of sufficient training data, the correct definition of "fair," and even the diversity of the professionals working in this area, are presented as problems.

It is impossible, for example, to measure the accuracy of an AI program that is developed by white heterosexual men if applied to other groups and assume it has no bias. These developers, even if they have vast knowledge of the world and its different situations, do not have the same experiences as people from other racial, ethical, and diverse groups, which is essential for an accurate measurement of the system. Which leads to the next issue: the need to comprehensively test the ML algorithm with different individuals based on real-life situations to ensure it is as accurate as possible.

However, we do know that this is rarely done due to both time and cost issues. This could mean, for example, that the machine has a certain bias toward an ethnic group simply because the data it was trained on had more data points pointing to them. While this will be beneficial for some groups, it might be highly prejudicial to others, even if that is not the intention. It is important to remember that sometimes this bias is not "intentional" but rather a result of the data available in the market.

"Bias may not be deliberate. It may be unavoidable because of the way that measurements are made—but it means that we must estimate the error (confidence intervals) around each data point to interpret the results" (Siwicki, 2021). Due to this, and added to the incapacity of a machine to make "rational" and "emotionally-oriented" decisions regarding the specific nuances where it is applied, there is a significant risk in using AI in certain fields. The implications of this can range from not being selected for a job, such as the Amazon case, to being incorrectly identified as an individual who committed a crime, as happened with police surveillance software. Let's take a closer look into these implications and how they can significantly impact our lives.

Implications

In 2018, it was announced that the ecommerce giant Amazon was going to recalibrate its hiring algorithm to better reflect the situation of the candidates. According to what was reported at the time, the system was biased against women and individuals who did not have certain keywords in their CVs. This meant that several female candidates who could have participated in its hiring process were discarded, since male candidates were "preferred" by what the algorithm had taught itself. "It penalized resumes that included the word 'women's,' as in 'women's chess club captain.' And it downgraded graduates of two all-women's colleges, according to people familiar with the matter. They did not specify the names of the schools" (Lauret, 2019).

In other situations mentioned by Best (2021), "Studies have found mortgage algorithms charging Black and Latinx borrowers higher interest rates. A series of studies about various facial recognition software found that most had misidentified darker-skinned women 37% more often than those with lighter-skin tones." These implications become even more significant if you consider that, in 2022, 48% of the CIOs in top companies announced that they would be launching and deploying an AI program within their organizations, or had already done it (Kaminska, 2022).

A clear example of AI bias was reported in a *Scientific American* article which identified AI bias against Black people in law enforcement agencies. "This results from factors that include the lack of Black faces in the algorithms' training datasets, a belief that these programs are infallible and a tendency of officers' own biases to magnify these issues" (Johnson and Johnson, 2023). This led to a discussion about the ethical implications of AI, racial profiling, and how technology can be misused in these cases when it comes to determining whether an individual will be arrested or not.

The matter is deeply analyzed by Kaminska (2022), who mentions a survey made by the company DataRobot which concluded that,

> According to the survey analysis, 54% of technology leaders say they are very or extremely concerned about AI bias. It's 12% more than in 2019 (42% shared such a sentiment). At the same time, the overwhelming majority (81%) are calling for more AI regulation. The concern over bias in AI is justifiable. Out of 350 organizations surveyed, 36% (126 organizations) said their organization suffered from biased data in one or several of their algorithms. (AI & Data Bias Concerns section)

You might be wondering, *If AI is based on data collected by humans and humans are naturally biased, is there a way to mitigate this bias or reduce it so the software is more reliable?* That is a great question that is at the center of discussion in today's AI market. Many companies and professionals, as you have seen by the numbers, are engaged in finding solutions to reduce its occurrence. Let's see some of the ways to make this happen.

Strategies to Mitigate AI Bias

Several different strategies can be used to mitigate AI bias, although it is nearly impossible to eliminate it completely. While there are practical applications such as listening to feedback, ensuring you have comprehensive data that comprise different aspects of the situation, and constantly performing quality checks to ensure that the machine is performing to standard, other changes need to be made at a structural level. These changes are more difficult to carry out and require a change in the market mentality that starts at the hiring process.

Companies who are developing AI software need to diversify their teams so that as many different points of view as possible are taken into consideration. This would mean hiring people from different educational institutions, for example, and people from diverse ethnic, racial, gender, age, and other groups. In addition to this, it is important to create an awareness that AI software is not perfect.

This means, for example, that individuals should understand that we are talking about a machine that does not make decisions based on all the different consciousness levels that humans use. They function based on the data they were fed and have their limitations regarding sensitivity and other restraints. This includes understanding how feedback is given, for example, and how this is affecting the way the AI performs, as well as the different options to mitigate this.

Other aspects such as algorithm transparency, laws and regulations, and even establishing parameters and specifications for these types of programs are important. Limiting or considering the purpose for which it will be applied should also be taken into consideration, as it should reflect the use the software will have. Manyika et al. (2019) mention five actions that could efficiently help in this process:

- Have company management and C-suite level executives keep up with the changing markets and the evolution of research in the field.

- Involve those who usually suffer from bias in the discussions of the creation of these programs and the different aspects that should be considered.

- Invest in generating and using data that has higher quality and that is more reliable, reflecting the diverse population that should be represented.

- Establish regulations and processes that will help mitigate AI bias from the start, from data collection to its deployment.

- Involve humans in all parts of the process to ensure that the bias is as minimal as possible and carry out comprehensive testing to ensure that bias is mitigated.

Essentially speaking, the task of mitigating bias in AI should be present from the beginning of the process— which is not the identification of the business problem. It is about the professionals who will be chosen and, subsequently, all the other steps that are needed to develop an AI software. This includes but is not limited to data collection, data preparation, machine training and testing, model selection, processing decisions, and even how the output will be presented to the final user.

It is possible to make AI fairer and less biased. Once companies understand this difficulty and the positive results from doing this, the use of AI software will be significantly more representative of the diverse society we live in. This does not mean it will be easy, and it will be neither cheap nor immediate. It is part of the process, one that has been set in motion and is ongoing but that still needs to be improved nonetheless. And this does not mean it is the only challenge AI users and developers face.

While an impressive technological development, AI is not devoid of flaws. Bias in AI can have significant real-world consequences, potentially perpetuating harmful stereotypes and discrimination. However, this is just one of many ethical considerations that come into play when we talk about AI. There are several other "gray areas" that should be considered, and this is exactly what we are going to look into now.

Chapter 7:

Navigating the Gray Areas

In January 2023, a study published by Study.com revealed that 89% of students used ChatGPT to do their homework or school tasks. In addition to this, 53% submitted an essay written by the tool, 48% admitted to using it for an at-home test or quiz, and 22% used it as a tool for writing the outline for a school paper. These statistics only refer to the educational area, but if it were to be expanded to businesses, the numbers are just as relevant and alarming.

This begs the questions, *What are the ethical limits to using AI tools for work or school-related activities? What are the regulations being applied to this technology to ensure its proper use? How can we ensure that AI is being used within ethical boundaries of what is acceptable?* These are all excellent questions, since AI is not just about technology; it's also about ethics.

AI has undoubtedly changed the way we work and use technology. It has affected different areas of our lives— from using it to create recipes with what we have in our refrigerators to helping create outlines for marketing campaigns. But what is the limit? Where should the line be drawn between what a machine should and should not do, especially if we think about what we have seen in the previous chapter regarding bias and sensitivity?

These are all good questions that must be debated and considered when we think about using AI technology in our lives. In this chapter, we'll navigate the complex landscape of AI ethics and regulations.

We will discuss the different aspects of AI and the "gray areas," which still have no effective control—especially in the business realm. You will understand the broader ethical implications of AI and learn how to navigate its regulations. You will see different cases of AI use that have led to questioning where and *when* the line will be drawn.

However, it is not only about understanding the limits, it is also about how *you* can ethically use AI and adopt and create responsible AI solutions. It is time that we start thinking about the impacts of technology in our lives and see how we can collaborate as *creators* rather than just *users*. By the end of this chapter, you will be better equipped to understand AI regulations and uses, as well as how you can introduce it to your business within an ethical framework.

Ethical Implications

In May 2023, a New York lawyer was caught by the court after using ChatGPT to write his petition for the case. While this may not have been a problem, it turned into international news when it was discovered that the AI software had used for the petition six "fake" cases that never actually happened when mentioning previous court decisions. According to the reports about the case, in which the client was suing an airline, "His legal team submitted a brief that cited several previous court cases in an attempt to prove, using precedent, why the case should move forward" (Armstrong, 2023).

When researching the cases, it was discovered that they did not exist. Not only was it identified that the lawyer used ChatGPT to make the case but also that it was allegedly not him who wrote it. While the lawyer was fined and disciplined, this brought up concerns of the potential risks and ethical use of AI in different aspects of our lives today.

Despite already knowing that bias plays an important role in the performance of these software and the answers they will give, there are many other ethical implications that should be considered for its use. We have already mentioned the bias examples in the police AI face-recognition software and the Amazon hiring practice, and there are several other situations in which there are ethical considerations, apart from the students and the New York lawyer. This brings us to an urgent matter that must be discussed among the authorities, technology companies, and even the population, since it is unlikely AI is going anywhere.

In a paper published in 2020, Ryan and Stahl debate the ethical guidelines of using AI. As you can see, the discussion had already started way before the Microsoft-powered tool came into use. In the research, Ryan and Stahl establish guidelines and points of concern that should be considered during the ethical development and use of AI, and these include eleven main points and their subtopics:

- **Transparency:** including the concepts of not only the first but also explainability, understandability, interpretability, communication, disclosure, and showing. Ask yourself the question, *How is this information being obtained, and where is this data the machine is learning sourced from?*

- **Justice and fairness:** together with the subcategories of consistency, inclusion, equality, equity, non-bias, non-discrimination, diversity, plurality, accessibility, reversibility, remedy, redress, challenge, and access and distribution. Ask yourself the question, *Is the information contained in this AI software just and fair for all the involved parties, or does it contain bias and partial information?*

- **Non-maleficence:** including the topics of security, safety, harm, protection, precaution, prevention, integrity, and non-subversion. Ask yourself the question, *Can this AI software bring any harm, especially for those who might think it can be used to replace official diagnosis, for example?*

- **Responsibility:** that would include items such as accountability, liability, and acting with integrity. Ask yourself the question, *Who is responsible if there are any negative repercussions of the obtained information?*

- **Privacy:** distinguishing between personal and private information. Ask yourself the question, *Is this information copyrighted and protected, or is it of free access?*

- **Beneficence:** with topics such as benefits, well-being, peace, and social and common good. Ask yourself the question, *Does this software provide negative information, or does it contribute to the well-being of the population and its users?*

- **Freedom and autonomy:** including the concepts of consent, choice, self-determination, liberty, and empowerment. Ask yourself the question, *Have those being mentioned by the software given authorization to use their names and liberty of requesting non-publication?*

- **Trust and trustworthiness.** Ask yourself the question, *Can the information in this software be trusted?*

- **Sustainability:** with matters concerning the environment, energy, nature, and resources. Ask yourself the question, *What are the energy resources and harm all this energy, processing, and electronic waste are generating to an already fragile planet?*

- **Dignity:** Ask yourself the question, *Does this software harm any individual's dignity with the information it contains or discloses?*

- **Solidarity:** mentioning topics like solidarity, social security, and cohesion. Ask yourself the question, *Who benefits from this information, and how are they compensated?*

Despite being a comprehensive list that makes it impossible to elaborate on all of the items in just one section of a book, there are certainly some pertinent points and questions that should be asked, as you have seen. The matter of using ethical AI tools is urgent. Companies need to consider if not all than most of the above to ensure that they are using the tools in the best way possible. For the good news, some advances have been made: If you enter any biased, racist, or similar content into ChatGPT, it essentially states that this is not authorized and that you should check your view of the world.

On the other hand, it still gives "advice" regarding diseases and suggestions to individuals on several different matters. In addition to this, to talk about other AI applications, there is a debate surrounding the data used to train these machines. If the machine does not "create" but rather learns from what it is fed, how can it "copy" art from illustrators and designers? Is this ethical? To train a machine on what someone else does, in their profession, so that it can generate this content?

The truth is that there is still much to be debated regarding the matter—especially if you consider the incredible quantity of lawsuits that are starting to be filed against these programs being taught with proprietary work. But we are going to talk about this, and more specifically legislation, in a little bit. However, before we get there, there is another matter that should also be discussed.

It is likely that you have heard about professionals who claim their job has been "stolen" by AI or that they have been replaced by AI. If you have, or even if you are one of these individuals, you are not alone. The perceived threat that AI has brought to several professionals is *real,* and it has been creating significant worries to certain segments of the population. If you are concerned about a potential "AI takeover" of your job, or the job of a loved one, let's take a look into what this *really* means and what can be expected in the future.

AI and Job Displacement

In June 2023, 24% of American workers claimed they were scared that AI would make their jobs obsolete. This worry was higher among those working in marketing and advertising (51%), business support and logistics (46%), and individuals who are Black (32%), Hispanic (35%), and Asian (38%). At the same time, "A majority (64%) of workers say they don't use AI at all on the job; 26% say AI can help but isn't necessary, and only 8% say using AI is required" (Rosenbaum, 2023). Finally, 43% of the interviewees claim they are aware that AI will likely change how their jobs are done in the near future.

To make a better argument, I want you to think about a moment in history when it seemed that everything was going to change, like the Industrial Revolution from

1760 to 1840. Before it happened, people lived off trade, sustenance agriculture, and small businesses. However, all this changed with the introduction of machines, factories, and later, energy. At the same time, we don't even need to go that far; there have been several different "innovations" throughout the last century that have disrupted jobs as they were carried out.

We could mention many of those whose jobs were changed, like those who lit the light posts at night before electric energy, telephone operators when the device was invented, and even the postman with the invention of the email. While the AI revolution is more connected to the Industrial Revolution than to these other situations, in which it affected not just a small part of the population but several industries as they were known, looking into the specific changes also helps shed some light on the matter. This is because these are exactly all the different aspects of work that we know will change.

Granted, there is no need to get into much detail, since you have already read in this book about all the different uses of AI and how it can be used in different industries. However, you have also read about the situations in which humans can simply not be replaced—not now and not in the near future. When the debate over AI as a job destroyer or a job creator emerges, I like to think that it can be a little bit of both, but it is more in the middle. It is similar, in my point of view, to how the Industrial Revolution changed the way society saw work and how these tasks were carried out, but at the same time that many jobs were lost, many others were created.

If you think about it, you will realize that we are exactly in the same place as in the mid-1700s: We are facing a revolution that will likely take away some jobs but also create the opportunity for the development of several others. While it is possible that a number of customer service agents will be replaced by chatbots with NLP, and that AI can make the lives of developers easier when it comes to writing code, AI also presents several opportunities. There will be an increase in the number of opportunities for those interested in ML and new areas of expertise, such as prompt engineering and even quality assurance for AI programs. These are all jobs that didn't exist before and have become a new need in the market.

At the same time, you must think that a writer, for example, will always be a writer. A machine will never be able to replace the creativity of a person or replace them when telling a personal story. A musician's fans will always want to see a live show, and this is something that AI cannot do. A doctor cannot be replaced, and a personal trainer who corrects the position of a student can't, either. AI cannot perform the CPR routine a lifeguard would perform, and AI cannot compete in the Olympics or any other sport that we watch for entertainment. (At least not yet.)

While we have several different applications for AI that can replace logical decision-making actions, it will never be able to replace a creative, innovative, or unique professional. Furthermore, even if you think about how work is constantly changing, remember that some of the jobs that existed when our grandparents were young do not exist anymore, though new ones were created. The same can be said for our parents' time and when our children grow up. The world is constantly changing, and AI is just another transformation that we need to understand and accept as we learn how to incorporate it into our lives and make the best use of it.

In the middle of 2023, the United States entertainment industry suffered the second longest stall ever in its production because of the Writers Guild of America (WGA) strike. Among the several claims made, there was a significant one that mentioned the use of AI in the entertainment industry. Professionals were worried that AI was going to take the writers' jobs and that the staff would be even more reduced with the help of AI tools. Furthermore, they were concerned that their previous work would be used to train the machines that would eventually replace them.

After many months of discussion and negotiation, an agreement was reached with the professionals that said they would not be replaced by AI and would have some job security in this regard. This was considered a victory, especially to safeguard the jobs of people who, at some point, could be replaced by technology. However, if you have ever tried writing an original text with feeling and sentiment using an AI tool, you will see that, so far, it is still far from the mark.

Based on this instability and the doubts that have arisen in the market about the legal use of AI, new regulations are being written and discussions are being carried out. Right at this moment, in different government and judicial spheres in several countries, the use of AI is being debated and legislators are thinking about how new regulations can be used to put some limits on the market. The European Union (EU) was a pioneer in this subject by starting to regulate how customers' data and information was being used and for what purposes.

Let's take a look into what we can expect regarding new legislation, how some countries are reacting, and other points that are still not being discussed but must also be considered.

Navigating AI Regulations

Despite not being necessarily a "new" technology, AI has increased its presence as a main character in our lives, and authorities have been moving to ensure that it is regulated. Countries like Brazil, Canada, Switzerland, India, China, and the United States have started drafting legislation pertaining to the matter, some of which are already being enforced, while others are still being reviewed. The fact is that the EU was a pioneer with what is called the AI Act, which "promotes investment and innovation in AI, enhances governance and effective enforcement of existing law on fundamental rights and safety, and facilitates the development of a single market for AI applications" (Council of the EU, 2022).

In this Act, the EU council establishes the practices that are prohibited, the classifications of high-risk AI systems and how they should be managed, the use of general purpose AI, and even transparency actions and what measures will be supported to increase innovation. Most of the premises deal with the matter of protecting the use of proprietary information, ensuring user safety, data ownership, and even the right to privacy of information.

At the same time, companies that develop AI software are currently facing legal action regarding the protection of user data. In the most notable case, OpenAI, the company behind ChatGPT, is facing a class action lawsuit that claims the company stole data from millions of users to train the algorithm. "The nearly 160-page complaint alleges that this personal data, including 'essentially every piece of data exchanged on the internet it could take,' was also seized by the company without notice, consent or 'just compensation'" (Thorbecke, 2023). As of the writing of this book, there has still been no decision on the matter.

Nevertheless, it is safe to state that there will be challenges faced by those who are looking to regulate the industry. One of the biggest is the speed with which changes are taking place while innovation is trailblazing. While legislation can take months, or even years, to put in place, technology is changing by the day, and it seems as if there is a new innovation every time we listen to the news. This might be one of the most difficult challenges to solve, since it requires an extensive review in the different legislative systems of countries and a constant adaptation to what is being released.

Next, legislators need to identify, when laws are passed, who will enforce this control and how. Assigning a responsible organization or public office to do this means investment in and hiring of those with expertise. Even if this is possible, there would need to be an exact definition of what is going to be regulated. Are we talking about regulating data? Programs? Companies? With the amplitude and reach of the internet, it is safe to say that it is unlikely that any of the solutions that have been proposed so far have managed to bring useful insights.

If an individual is using AI to break a bank program and steal money, who will be accountable? The individual who put the software to work, or the one who created the algorithm? Some might even say they did not do anything wrong, since it was the computer that was at fault. There are still many layers that need to be uncovered and more problems than solutions at this point. This means that today, the ethical and responsible use of AI depends on us, the users, and how we apply it to our lives.

Developing Responsible AI Solutions

The way we use AI today will certainly be different from how AI will be used by future generations. One of the main factors that will influence this is exactly what we teach our youth about the ethical and responsible uses of technology. I would say that, more than what they will be taught in schools or by society, it will depend on what adults show these youngsters that will determine how it will be applied—for better or worse.

But what about today? What can be done today to use AI responsibly and within ethical and appropriate limits?

An organization that wants to responsibly implement AI within their hiring system might want to consider embedding "responsible AI principles into its AI governance, policies, and practices and take steps to understand and mitigate risks" (Lawson, 2023). Each company, as you navigate their website or even carry out a search in your browser, is implementing these solutions based on their profiles, needs, and use. Most of them consider the transparent, fair, and safe approach to the use of the tool. Examples include companies such as Accenture, Google, Amazon, Microsoft, and Apple (Jackson, 2023), with the number growing as time goes on.

As of now, it is unlikely that a company will look into all of the AI ethical matters that we mentioned in the beginning of this chapter. Nonetheless, the public is starting to ask for action, and customers are looking into how these organizations are using their data and that of others. It was even suggested, as a way of regulating the industry, that an AI certification system be established to help audit companies regarding their AI use. At the moment, all these solutions are being discussed and attentively followed by users—but it still remains to be seen what will be done in the future.

We have now journeyed through the complex world of AI ethics and regulations, shedding light on the crucial need for responsible AI solutions. However, as we continue to navigate the ethical and regulatory challenges, AI itself continues to evolve at a staggering pace.

Chapter 8:

Looking Ahead

AI is not static; it's continuously evolving. As you have seen in the previous chapter, this quick transformation and adaptation is one of the main challenges faced by different organisms to enforce regulation. *But what is the future preparing for us?* That is an excellent question, and one we can only begin to imagine: self-driving cars, humanoid robots, and the increase in efficiency of what we have today, to name only a few.

As we stand on the brink of a new era, this chapter will guide you through the future trends and advancements in AI. You will understand the future of AI, including emerging trends like Generative Adversarial Networks (GANs) and transfer learning. You will read about some of the technology that is being developed and how it has the potential to change our lives. You will learn what is being tested and implemented in different industries so that you might know what to expect to see in the future.

This final chapter is all about looking forward with an eye on the present. If you are ready to take this last deep dive into what the future holds, join me as we finish this last part of our journey through the AI world. By the end of this chapter, you will be able to keep pace with AI advancements and stay updated in the field.

Emerging Trends in AI

As the AI industry grows, it is forecasted that investments in the area should reach $15.7 trillion, a number 14% higher than what it is today (Analytics Vidhya, 2023). This means it is likely that we are set to see even more disruptions than what we have seen so far in the market in different industries and with distinct applications. We already know the power the technology has to change our lives and even improve some aspects of it (If you haven't tried asking a program what recipe you can make with only what you have in the fridge, you should try it!).

One of the most popular trends that is likely to grow is the use of NLP in chatbots and applications that will help make communication easier. Whether we are talking about facilitating healthcare options for disabled patients, providing 24-hour customer service for a website, or even simultaneous translations for individuals speaking different languages, there is a high probability that we will discover new ways to communicate. In addition to this, we also have features which are already being used, such as transcribing audio to text, summarizing and analyzing videos and documents, and even identifying music.

At the same time, while Microsoft has already integrated AI into its browser, Bing, embedding AI tools with other search engines is likely to happen sooner rather than later. While some software is already able to detect what a certain web page is saying, soon it won't take as much effort to find something based on our specific needs. "Instead of producing a list of relevant websites, the tool gives written answers that were pulled from a combination of different resources. Users can query broad questions like how to plan a three-course meal or which car to buy" (Howarth, 2023).

Of course, we could not leave out the potential developments that AI is bringing to the market, such as the often-mentioned autonomous vehicle, which will use different AI tools such as deep learning and computer vision. The car that does not need a driver has long been in talks in the market, and there are several companies working on its creation, most notably the Elon Musk-spearheaded Tesla.

Other developments include AI-based software that helps students with their profiles and applications to get into their desired universities, the use of machine-trained algorithms that can detect cancer in image exams that might be missed by the human eye, and even helping teachers identify the best tutoring methods to help students who need extra material to study. According to Howarth (2023), "Adoption among hospitals is surging—90% of hospitals have an AI strategy and 75% of hospital executives say AI initiatives are critical."

Finally, we need to mention the application of AI in commerce and ecommerce in general. When "dynamic" pricing was introduced to the market in 1980 by airline companies, and which became widely known with Uber starting in 2011, many people did not understand the "fairness" of the process or how this was determined. Despite understanding the concept of higher demand and less supply leading to higher prices, few people knew that the company used an AI application for this process. Today, more online platforms selling products are using this concept to ensure they have the best yield. AI solutions are not all alike—far from it—so who knows if you might find one for your own business soon?

Generative Adversarial Networks and Transfer Learning

We got a comprehensive overview of the developments in AI for several industries in the previous chapter, but there are two "new" concepts concerning AI development that are currently trending and earning the highest volume of investments. These are known as generative adversarial networks (GANs) and transfer learning. While the first, GANs, will generate new artificial, or synthetic, data that seems real, transfer learning means exactly what you might imagine. It is the process of reusing already developed, tested, and working ML algorithms for other purposes.

In this section of the chapter, we will take a deeper look into these two types of AI and how they are already impacting the way we use the technology. Once you better understand what these are, I am certain you will identify at least one application where they have been applied or seen an example of these in action. As you will be able to see, GANs and transfer learning are incredibly powerful tools that have the ability to change how we see the world—literally.

Generative Adversarial Networks

GANs, as they are known, have an apparently complicated name, but its application is actually pretty straightforward. Suppose that you have a picture of yourself and you would like to see what you would look like with your face on the body of a superhero. Interesting, right? Well, this is what the GANs will enable you to do, since their main purpose is to use real information and, based on this, create artificial images that make it almost impossible to identify whether it is real or not. This means that if you *wanted* someone you don't know to believe that you have the body of a superhero, it is possible to make this happen. However, how ethical is this?

Before we get into the debate, uses, and challenges of GANs, we should start out by understanding what these are. GANs are ANNs that use deep learning to train the machine. In this case, the approach usually used by specialists is the unsupervised ML model "that involves automatically discovering and learning the regularities or patterns in input data in such a way that the model can be used to generate or output new examples that plausibly could have been drawn from the original dataset" (Brownlee, 2019).

In this case, the definition will help us understand why this is called a *generative* adversarial network. If you review the contents of Chapters 1 and 2, when we talked about the different AI and ML types, you will remember that when we refer to unsupervised learning, there is no expected output. Therefore, this means the ML process will determine what the outcome will be, and then the example will be evaluated to ensure that it has the right parameters.

This is exactly the case with the GANs. In this case, it uses the DL of ANNs to create new variables that might fit into the "problem" that has been given to the machine. In our case, if we are talking about changing a picture to the body of a superhero, this means matching the lighting on the image, backgrounds, skin color, and much more to ensure that it is as perfect as possible. In this case, we could say that "the GAN model architecture involves two sub-models: a *generator* model for generating new examples and a *discriminator* model for classifying whether generated examples are real, from the domain, or fake, generated by the generator mode" (Brownlee, 2019).

When you consider its capability, it is possible to imagine a wide range of applications for this tool, many of which you have probably already seen:

- filters that show you how you will look when you have aged

- transforming pictures into cartoons

- using written prompts to generate original images

- improving photo resolution

- blending two pictures together

- general uses of photographic and image editing

- generating image datasets to train other AI algorithms without needing to use images of "real" people

- recovering damaged photographs and videos

- creating deep fakes that are almost impossible to identify

As you might imagine, this technology raises several ethical problems, some of which have already been mentioned, such as: Where is the data being obtained from? Is there authorization to use this data? What is the impact on society with the use of deep fakes? For the time being, companies are still investing and looking into the best alternatives to develop and enhance this process, since there is still a learning curve. It is safe to say that today, GANs present the most "accurate" results when considering real images, a matter that can also be expanded to pictures, videos, music, data expansion and prediction, and so much more.

However, it is not without its difficulties. Some of the challenges faced by developers, mentioned by Roy (2023) include the instability that these networks present while being trained with a large variation of the output, overfitting the data and making the output too similar to the original images that were used to train it, the transparency in explaining how these GANs work and their learning process, and even the cost that it takes to train a machine like this, considering the hardware and the required knowledge to do so, which can be expensive. Finally, we should mention that there is also the matter of bias and fairness, since the synthetic result can prove to be discriminative depending on what was used to train it.

One of the alternatives that are being looked into that could help solve, at least partially, some of these issues with GANs are exactly what we are about to dig into next: transfer learning. What do you think would happen if we, humans, had a process that enabled us to learn from our parents in a way that was as simple as reallocating the knowledge? This could mean that many people would not need to undergo the process of learning how to walk, for example. Interesting, right? Let's see how this works.

Transfer Learning

Although the previous section teased about humans learning from others, this is actually what happens today, if you think about it. We go to school to learn from our teachers. We learn manners and how to behave from our parents. Our friends teach us about new music and introduce us to things we might like. All of this is a learning process and can be considered a transfer of knowledge. However, it takes years, even decades, for this process to be accomplished, and we are still learning throughout life.

In the case of machines, this is done automatically by getting one model and testing to see if it fits another purpose—almost as easy as pushing a button and "downloading" all the knowledge a college teacher has into the head of a toddler, for an exaggerated example. In this case, we could say that a machine algorithm that has been already trained can be used for the application in a new task or new process, thus minimizing the costs of development and (potentially) the time it would take to become productive.

The concept is rather simple and based on an article written by Lobo in 2017; as you can see by the date, it has been in place for some time. *But what has changed?* you might ask. In this case, we need to look into the different concepts of how AI was used in the recent past and how it is being used today. This means that while we previously had AI for translations, recommendation systems, GPS, and virtual assistants, if you look at the quality of what we have today, it has significantly increased. This means that the AI software and ML processes that we have today are more developed than what we had before and, thus, it makes more sense to use them to transfer the application rather than developing from scratch.

If we think about its application in computer vision, for example, we could say that the same software that was used to analyze football games was adapted and can now carry out a similar analysis for basketball, volleyball, and other sports. The same can be said about using NLP; some programs that have used ML processes to adapt the NLP for a chatbot for a specific industry can be adapted to others with less training, since it already knows the optimal parameters. Finally, if we consider image changes, it would be possible to say that if a computer is able to predict what an elephant will look like, it can easily be trained to identify hippos.

I bet that by now, you understand how and why each of these two tendencies in ML and AI are so intertwined—the possibilities of a machine that already has optimal performance being trained to fix the challenges faced with GANs is incredible. Not only this, but it can significantly expedite the process of making AI a more accessible tool for small and big companies alike, enabling different businesses to also participate in the technological advances of the sector.

It still remains to be seen how this will work out and if there will be any success. While developers have been able to successfully carry out transfer learning for similar purposes, its application for distinct processes will still take some time. However, it is not impossible to imagine that if we are able to transfer learning between different machines and build one with all the features, what we will have is a next step in the AI revolution, including bringing us closer to the theory of mind and self-awareness in AI that you have seen in movies.

To ensure that you have the correct tools to understand and keep up with the latest tendencies in the market, we will now present two lists. The first is a list of resources that can help you stay informed, and the second is a list of tools that are already available and can be used by accessing the programs' websites. I encourage you to take note of these or even navigate through them as you read so you can take note of the ones that interest you the most for further investigation.

Staying Up-To-Date With AI Advancements

As we conclude our exploration into the intricate and thrilling world of artificial intelligence, it's clear that we are just at the cusp of its potential. AI, with its vast applications and continuous advancements, is undeniably transforming our world. However, staying up to date with this rapidly evolving technology can be as challenging as it is exciting. As we move forward, it is safe to say that continuous learning and adapting will be key to "surviving" in an AI-driven world.

Regardless of whether you are a university student just entering the world of technology, a member or manager of an IT team, or even part of the C-suite in a company: AI will definitely change the way you live and interact with technology. While for those who are entering or studying in a university, this can be a path to a new and exciting career, those in a business's top management may use it to leverage its potential. For those who are already in the IT industry, it could mean a change in career path or a way to give the department a new "shine."

But these are not the only organizational areas that will be affected. AI concerns all of us: business and data analysts, restaurant chefs, customer service departments, product development teams, and even human resources. At the same time, you might imagine that AI will *never* be used in the field that you work in, but you never know with the speed of the new developments.

Therefore, instead of getting caught by surprise, the best course of action is to keep up with the news of the market so that you are updated on the ongoing developments. You don't need to check in every day, but sometimes just skimming around to look and see if there is anything that can be applied to you or to your life is beneficial. This is what this final section of the chapter, and of the book, is all about.

The following sections will be all about different AI tools and what they do, as well as resources that will help you learn more. These are sources such as magazines and blogs that will help you keep yourself updated and tools that will show you the different things AI can do. Get ready to discover some of the most amazing tools and the best places to learn more about AI!

Resources to Learn More

Considering that the internet sometimes has information of doubtful quality and credibility, it is important to know where to get educated. This is the case when you want resources that can give you the ability to understand more about what AI advancements are taking place and how they are being applied. Here are a few sources that you might want to consider checking out if you want (and you should!) to stay informed about AI:

- AI Coffee Break with Letitia (YouTube channel)

- AI Magazine

- AI News

- AI Scout (website)

- Analytics Insight (magazine and website)

- Data Science Weekly (newsletter)

- DeepMind Blog (blog)

- Extreme Tech (website)

- Google AI blog (blog)

- Great Learning (website)

- KDNuggets (website)

- MIT News (website)

- OpenAI blog (blog)

- Towards Data Science (website)

- Towards AI (website)

- Wired (website)

Most of the resources named in this section have more general information than specific technical material that can be used if you are a developer. Nonetheless, some of them still have sections like tools and articles on how to enhance your ML process and help with generating AI software. The second type of articles are more likely to have ML and AI jargon, so if you are more comfortable avoiding these terms, these resources are a good place to start.

In addition to this, despite mentioning the blog for a few specific companies, there are several others that have specific AI content on their websites that can help you understand more about a specific industry. Some of them include Meta, Microsoft, and Apple. If you work for a more specific industry, such as healthcare, pharmaceuticals, or even sports, there might be other pages that you can be directed to that will ensure you have the adequate information. In this case, to ensure they are reliable, remember to check the date of what you are reading, the author, and the credibility of the website. While I would greatly wish to name all the amazing places there are to obtain information from, it would be a rather large list to account for every specific industry!

This being said, I can bet that you are anxious to learn more about some of the AI tools that can be used and are not commonly mentioned in the "market buzz." For this reason, in the next and final section of this chapter, I will provide you with some of these tools and software you can check out and test. Most of them have a free feature that does not require payment, although it is likely you will need to log in to some of them with your credentials for registration purposes. Read on to find out what the AI world is already working on and see if there is anything (within ethical boundaries) that you can use!

Tools to Try

Listed below are 20 different AI tools for you to try that are based on my experience and on the lists provided by Rebelo (2023), McFarland (2023), and Parker (2023):

- **Anyword:** AI software that is being used by those in the marketing and publicity area to help create content and generate outlines for how it should be written.

- **Aomni:** AI software that aids in researching scientific topics and returns a summary of the results.

- **Arigram:** Used to transcribe audio to text in an advanced manner. Similar functions can be seen in Microsoft programs, such as Word, that also have this feature for different languages.

- **Bard:** Google's chatbot with NLP, equivalent to ChatGPT.

- **Chatfuel AI:** Created to help companies develop and configure their own chatbots for customer service that can be applied to business websites.

- **Cody by Sourcegraph:** Can be used by developers to help optimize, fix, understand, and even create and automate code for software.

- **DALL·E 2:** An image generating software that creates images based on prompts and descriptions given by the user.

- **EmailTree:** Helps you organize your email inbox and control what needs to be done to specific messages.

- **Feathery:** Used to help clients create elaborate forms and surveys for data-gathering without needing to code.

- **HitPaw:** Helps with fixing pictures that are blurry, recovers quality, and enhances images.

- **Jasper:** An AI software that helps with content development and even with writing your own books!

- **Mem:** Is used to help people take notes, and the program will automatically connect the content for you.

- **Murf:** Text-speech generator used to create narrations and voice content based on text. Has an incredible database of voices and even allows you to modify your voice in recordings.

- **Neuraltext:** A content builder that puts together the features of keyword search and an SEO tool to create material that will be used for marketing and advertising. Has an additional feature that will "grade" the SEO content you create.

- **Paradox:** AI software used to optimize human resources processes. Can perform tasks that will help with the selection process from recruiting to onboarding.

- **PlusAI:** GAN tool that will help you generate custom slides in Google Slides.

- **ProWritingAid:** A text "helper" that will check grammar, spelling, style, and different aspects of what you are writing. It gives you suggestions and grades what you have written to help you enhance the text.

- **Reclaim:** Program used to help you organize your agenda and even changes and reallocates commitments to other times if you miss one.

- **Runway:** Used to edit, produce, and generate video that brings amazing features using GAN.

- **Slides:** Used to create presentations by telling it about the topic and the metrics you want to use. Will generate different types of slides, formats, and text and personalize the context.

As you can see, there are plenty of tools that can be used for the most varied purposes. Maybe applying one or more of these to your day will help you optimize some tasks and give you more time. If you take a look, they are not all necessarily used for business purposes: Agenda organizers can help mothers with children's schedules, text-to-speech can help the visually disabled read books that have not been transformed into audio books, you can improve and enhance pictures that were taken long ago and that need an improvement in quality, and even the enhanced grammar and spelling check can help you with your emails.

In the end, despite several AI programs presenting ethical dilemmas for the users and creating controversy, there are some tools that can help us without (apparently) breaking any legal boundaries. As we wrap up this final chapter and move on to the conclusion, I am certain that you have a lot to think about and consider, regardless of what you dedicate your life to. You have already taken the first step to making AI better by understanding its nuances and the challenges, and now it is up to you. Are you ready to be a part of the solution?

Conclusion

There is no doubt that AI is here to stay. It's always better to understand this technology rather than fearing a future that's far from happening. In this book, you have been through a journey that has given you a better understanding of what AI is, how it can be used, and the challenges it faces and brings us. You have learned why some AI software "behaves" the way it does and even how it is possible for a machine to have bias.

You are now duly equipped with all the necessary knowledge to engage in conversations that refer to the subject, its implications, dilemmas, and perspectives. This is a discussion that, as mentioned earlier, *needs* to be expanded to reach different spheres of society. It is essential that we have lawmakers, businesspeople, our community, teachers, doctors, and everyone else engaged. This is how we take our stand and demand that our rights, jobs, information, privacy, and even art are preserved. It must be discussed inside the classroom in middle schools, in universities, and in board rooms with decision-makers.

As you have read throughout this book, the changes are happening—and fast. In an intended pun, we can say that the machines are learning every minute of every day, and if we don't start taking action now, it might reach an irreversible point. Some people are already doing this by using their unions and their voices to speak up and speak out.

I can imagine that you are right now thinking about all the implications that AI will have in your life and the lives of your loved ones in the near future. Maybe you are thinking about how it will affect your business, the children in your family, and even society as we know it today. There is certainly a lot to think about, consider, and reflect upon, and this is made easier when you have all the necessary information.

In the final section of the last chapter of this book, you were given resources that can be used to keep updated about AI development. Use them! Don't let your interest fade away after you close this book or after discussing it over dinner. Keep connected to the trends in AI and the changes it is promoting in the different industries. Use the lessons you've learned from this book and start preparing for an AI-driven future.

As a last thought, if you feel that this book has helped you better understand the nuances of AI and what it means for today and for tomorrow, I encourage you to leave a review. Tell others how they will benefit from this book and the experience it has taken you through. As you know, the more people who are involved in the matter, the better the process will become. I wish you the best in your journey and hope that you have found the answers you were looking for in this book. Good luck!

References

AI investment forecast to approach $200 billion globally by 2025. (2023, August 1). Goldman Sachs. https://www.goldmansachs.com/intelligence/p ages/ai-investment-forecast-to-approach-200-billion-globally-by-2025.html

Akella, B. (2023, September 2). *Types of machine learning - supervised, unsupervised, and reinforcement.* Intellipaat. https://intellipaat.com/blog/tutorial/machine-learning-tutorial/types-of-machine-learning/?US#no2

Analytics Vidhya. (2023, September 5). *Top 20 trends in AI and ML to watch in 2023.* https://www.analyticsvidhya.com/blog/2023/0 5/emerging-trends-in-ai-and-machine-learning/#Emerging_Trends_in_AI_and_Mach ine_Learning

Anyoha, R. (2017, August 28). *The history of artificial intelligence.* Science in the News; Harvard University. https://sitn.hms.harvard.edu/flash/2017/histor y-artificial-intelligence/

Armstrong, K. (2023, May 27). *ChatGPT: US lawyer admits using AI for case research.* BBC News. https://www.bbc.com/news/world-us-canada-65735769

Best, M. (2021, June 29). *AI bias is personal for me. It should be for you, too.* PwC. https://www.pwc.com/us/en/tech-effect/ai-analytics/artificial-intelligence-bias.html

Betz, S. (2022, August 25). *4 types of artificial intelligence.* Built In. https://builtin.com/artificial-intelligence/types-of-artificial-intelligence

Biswal, A. (2023, May 26). *7 types of artificial intelligence that you should know in 2020.* Simplilearn. https://www.simplilearn.com/tutorials/artificial-intelligence-tutorial/types-of-artificial-intelligence

Brown, S. (2021, April 21). *Machine learning, explained.* MIT Sloan. https://mitsloan.mit.edu/ideas-made-to-matter/machine-learning-explained

Brownlee, J. (2019, June 16). *A gentle introduction to generative adversarial networks (GANS).* Machine Learning Mastery. https://machinelearningmastery.com/what-are-generative-adversarial-networks-gans/

Castillo, D. (2021, June 29). *Transfer learning for machine learning.* Seldon. https://www.seldon.io/transfer-learning

Coateswroth, B. J. (2023, April 13). *Unpacking the cybersecurity risks of artificial intelligence (AI).* LinkedIn. https://www.linkedin.com/pulse/unpacking-cybersecurity-risks-artificial-intelligence-coatesworth/

Cooban, A. (2023, July 23). *AI investment is booming. How much is hype?* CNN. https://edition.cnn.com/2023/07/23/business/ai-vc-investment-dot-com-bubble/index.html

Council of the EU. (2022, December 6). *Artificial Intelligence Act: Council calls for promoting safe AI that respects fundamental rights.* Consilium. https://www.consilium.europa.eu/en/press/press-releases/2022/12/06/artificial-intelligence-act-council-calls-for-promoting-safe-ai-that-respects-fundamental-rights/

Coursera. (2023, January 12). *4 types of AI: Getting to know artificial intelligence.* https://www.coursera.org/articles/types-of-ai

Delagrange, K. (2022, May 6). *Six misconceptions about artificial intelligence.* Spiria. https://www.spiria.com/en/blog/artificial-intelligence/6-misconceptions-about-artificial-intelligence/

Feature engineering. (n.d.). Heavy.AI. https://www.heavy.ai/technical-glossary/feature-engineering

4 types of machine learning (supervised, unsupervised, semi-supervised & reinforcement) - databasetown. (2020, January 6). Database Town. https://databasetown.com/types-of-machine-learning/

Frank, A. (2021, May 13). *Are we in an AI summer or AI winter?* Big Think. https://bigthink.com/13-8/are-we-in-an-ai-summer-or-ai-winter/

Fraudwatch. (2021, May 10). *How artificial intelligence is used for cyber security attacks.* FraudWatch. https://fraudwatch.com/how-artificial-intelligence-is-used-for-cyber-security-attacks/

Frawley, J. (2023, April 12). *3 things workers can do to prepare for AI and the future of work.* Fast Company. https://www.fastcompany.com/90879770/3-things-workers-can-do-to-prepare-for-ai-and-the-future-of-work

Generative AI. (n.d.-a). https://generativeai.net/

Generative AI. (n.d.-b). Accenture. https://www.accenture.com/us-en/insights/generative-ai

Generative AI: What is it, tools, models, applications and use cases. (n.d.). Gartner. https://www.gartner.com/en/topics/generative-ai

Georgian. (2017, August 25). *An overview of applied artificial intelligence.* https://georgian.io/applied-artificial-intelligence-overview/

Goergen, A. (2022, November 16). *What is transfer learning and why does it matter?* Levity.ai. https://levity.ai/blog/what-is-transfer-learning

Great Learning Team. (2023, August 18). *What is machine learning? Definition, types, applications, and more.* https://www.mygreatlearning.com/blog/what-is-machine-learning/

Harkiran78. (2023, June 2). *Artificial neural networks and its applications.* GeeksforGeeks. https://www.geeksforgeeks.org/artificial-neural-networks-and-its-applications/

Hintze, A. (2016, November 14). *Understanding the four types of artificial intelligence.* Govtech. https://www.govtech.com/computing/Underst

anding-the-Four-Types-of-Artificial-Intelligence.html

Hosanagar, K. (2020, January 21). *AI: A job killer or creator?* Business Today. https://www.businesstoday.in/specials/anniversary-special-2020/story/ai-a-job-killer-or-creator-243586-2020-01-21

Howarth, J. (2023, September 21). *7 top AI trends (2023 & 2024).* Exploding Topics. https://explodingtopics.com/blog/ai-trends

IBM. (2023). *What is deep learning?* https://www.ibm.com/topics/deep-learning

Ilija Mihajlovic. (2019, April 25). *Everything you ever wanted to know about computer vision.* Medium. https://towardsdatascience.com/everything-you-ever-wanted-to-know-about-computer-vision-heres-a-look-why-it-s-so-awesome-e8a58dfb641e

Jackson, A. (2023, July 12). *Top 10 companies with ethical AI practices.* AI Magazine. https://aimagazine.com/ai-strategy/top-10-companies-with-ethical-ai-practices

Jain, P. (2021, July 23). *Artificial intelligence in agriculture: Using modern day AI to solve traditional farming problems.* Analytics Vidhya. https://www.analyticsvidhya.com/blog/2020/1

1/artificial-intelligence-in-agriculture-using-
modern-day-ai-to-solve-traditional-farming-
problems/

Johnson, T., & Johnson, N. (2023, May 18). Police
facial recognition technology can't tell black
people apart. *Scientific American*.
https://www.scientificamerican.com/article/po
lice-facial-recognition-technology-cant-tell-
black-people-apart/

Kaminska, J. (2022, June 29). *The impact of data bias on
your business & the benefits of fair AI*. Statice.
https://www.statice.ai/post/data-bias-impact

Kidd, C., & Miller, A. (2023, October 21). *What is AI-
as-a-service? AIaaS explained*. BMC Blogs.
https://www.bmc.com/blogs/ai-as-a-service-
aiaas/

Kleinings, H. (2022, November 16). *What is AIaaS?
Your guide to AI as a service*. Levity.
https://levity.ai/blog/aiaas-guide

Kohn, B., & Pieper, F.-U. (2023, May 10). *AI regulation
around the world*. Taylor Wessing.
https://www.taylorwessing.com/en/interface/
2023/ai---are-we-getting-the-balance-between-
regulation-and-innovation-right/ai-regulation-
around-the-world

Korobeyko, T. (2023, January 25). *The ultimate list of machine learning statistics for 2023*. ITransition. https://www.itransition.com/machine-learning/statistics#

Larkin, Z. (2022, November 16). *AI bias - What is it and how to avoid it?* Levity. https://levity.ai/blog/ai-bias-how-to-avoid

Lauret, J. (2019, August 16). *Amazon's sexist AI recruiting tool: How did it go so wrong?* Becoming Human: Artificial Intelligence Magazine. https://becominghuman.ai/amazons-sexist-ai-recruiting-tool-how-did-it-go-so-wrong-e3d14816d98e

Lawson, A. (2023, January 24). *AI vs. responsible AI: Why it matters.* RAI Institute. https://www.responsible.ai/post/ai-vs-responsible-ai-why-is-it-important

Lobo, S. (2017, November 15). *5 cool ways transfer learning is being used today.* Packt Hub. https://hub.packtpub.com/5-cool-ways-transfer-learning-used-today/

Manyika, J., Silberg, J., & Presten, B. (2019, October 25). *What do we do about the biases in AI?* Harvard Business Review. https://hbr.org/2019/10/what-do-we-do-about-the-biases-in-ai

Market Trends. (2020, July 10). *The most common misconceptions about artificial intelligence to avoid.* Analytics Insight. https://www.analyticsinsight.net/the-most-common-misconceptions-about-ai-to-avoid/

Marr, B. (2021, July 2). *What is deep learning AI? A simple guide with 8 practical examples.* Bernard Marr. https://bernardmarr.com/what-is-deep-learning-ai-a-simple-guide-with-8-practical-examples/

Mathworks. (2019). *What is deep learning? | How it works, techniques & applications.* https://www.mathworks.com/discovery/deep-learning.html

McFarland, A. (2023, October 20). *10 "Best" AI tools for business (October 2023).* Unite.AI. https://www.unite.ai/best-ai-tools-for-business/

Medade, T. (2023, April 22). *Artificial intelligence (AI) has become a buzzword in recent years.* Medium. https://tushar-medade.medium.com/artificial-intelligence-ai-has-become-a-buzzword-in-recent-years-3c8792905c2d

Mirza, A. (2023, June 7). *Top 10 predictions for AI and emerging technology trends in 2024.* Medium. https://levelup.gitconnected.com/top-10-

predictions-for-ai-and-emerging-technology-
trends-in-2024-11b5f81f5d6c

Mittal, S. (2020, January 27). *What is applied AI or
artificial intelligence? Everything you need to know.*
Analytix   Labs.
https://www.analytixlabs.co.in/blog/what-is-
applied-ai/

Mixson, E. (2020, November 13). *What is applied artificial
intelligence (AI)?* AI, Data & Analytics Network.
https://www.aidataanalytics.network/data-
science-ai/articles/applied-ai-enterprise-data-
science

Moisset, S. (2023, May 24). *How security analysts can use
AI in cybersecurity.* FreeCodeCamp.
https://www.freecodecamp.org/news/how-to-
use-artificial-intelligence-in-
cybersecurity/#how-ai-is-used-in-cybersecurity

Muhd, M. N. (2022, February 14). *Applied AI - what it
really means in practice.* LinkedIn.
https://www.linkedin.com/pulse/applied-ai-
what-really-means-practice-m-nazri-muhd/

Ortiz, S. (2023, September 14). *What is generative AI and
why is it so popular? Here's everything you need to
know.* ZDNET.
https://www.zdnet.com/article/what-is-
generative-ai-and-why-is-it-so-popular-heres-
everything-you-need-to-know/

Osakwe, F. (2023, May 22). *The risks and rewards of artificial intelligence in cybersecurity*. Forbes. https://www.forbes.com/sites/forbestechcoun cil/2023/05/22/the-risks-and-rewards-of-artificial-intelligence-in-cybersecurity/?sh=14e9d62096db

Parker, H. (2023, October 2). *The always up-to-date list of the 50 best AI tools in 2023*. ClickUp. https://clickup.com/blog/ai-tools/

Patel, H. (2021, September 2). *What is feature engineering — Importance, tools and techniques for machine learning*. Medium. https://towardsdatascience.com/what-is-feature-engineering-importance-tools-and-techniques-for-machine-learning-2080b0269f10

Pazzanese, C. (2020, October 26). *Ethical concerns mount as AI takes bigger decision-making role*. Harvard Gazette; Harvard University. https://news.harvard.edu/gazette/story/2020/10/ethical-concerns-mount-as-ai-takes-bigger-decision-making-role/

Press, G. (2021, May 19). *114 milestones in the history of artificial intelligence (AI)*. Forbes. https://www.forbes.com/sites/gilpress/2021/05/19/114-milestones-in-the-history-of-artificial-intelligence-ai/?sh=2bc1a87b74bf

Raj, R. (n.d.). *Supervised, unsupervised, and semi-supervised learning with real-life usecase.* Enjoy Algorithms. https://www.enjoyalgorithms.com/blogs/super vised-unsupervised-and-semisupervised-learning

Rangaiah, M. (2021, March 8). *History of artificial intelligence with timeline.* Analytics Steps. https://www.analyticssteps.com/blogs/history-artificial-intelligence-ai

Rebelo, M. (2023, July 13). *The best AI productivity tools in 2023.* Zapier. https://zapier.com/blog/best-ai-productivity-tools/

Reidy, T. (2022, March 1). *AI and the future of work how to prepare for a job that doesn't exist yet.* LinkedIn. https://www.linkedin.com/pulse/ai-future-work-how-prepare-job-doesnt-exist-yet-tom-reidy/

Rocca, J. (2019, January 7). *Understanding generative adversarial networks (GANs).* Towards Data Science. https://towardsdatascience.com/understanding-generative-adversarial-networks-gans-cd6e4651a29

Rosenbaum, E. (2023, June 8). *These are the American workers most worried that AI. will soon make their jobs obsolete.* CNBC. https://www.cnbc.com/2023/06/08/these-are-

the-workers-most-worried-that-ai-will-soon-
take-their-jobs.html

Rosenbaum, E., & Anwah, O. (2023, June 23). *A.I. is now the biggest spend for nearly 50% of top tech executives across the economy: CNBC survey.* CNBC. https://www.cnbc.com/2023/06/23/the-ai-spending-boom-is-spreading-far-beyond-big-tech-companies.html

Roser, M. (2022, December 6). *The brief history of artificial intelligence: The world has changed fast – what might be next?* Our World in Data. https://ourworldindata.org/brief-history-of-ai

Rouse, M. (2023, June 27). *What is generative AI?* Techopedia. https://www.techopedia.com/definition/34633/generative-ai

Roy, R. (2023, June 10). *Generative Adversarial Network (GAN).* GeeksforGeeks. https://www.geeksforgeeks.org/generative-adversarial-network-gan/

Ryan, M., & Stahl, B. C. (2020). Artificial intelligence ethics guidelines for developers and users: clarifying their content and normative implications. *Journal of Information, Communication and Ethics in Society, 19*(1). https://doi.org/10.1108/jices-12-2019-0138

Sajid, H. (2023, April 1). *AI in cybersecurity: 5 crucial applications.* V7 Labs. https://www.v7labs.com/blog/ai-in-cybersecurity

Salian, I. (2018, August 2). *Supervised vs. unsupervised learning.* NVIDIA Blog. https://blogs.nvidia.com/blog/2018/08/02/supervised-unsupervised-learning/

Simplilearn. (2023, June 6). *What is data wrangling? Benefits, tools and skills.* https://www.simplilearn.com/data-wrangling-article

Siwicki, B. (2021, November 30). *How AI bias happens — and how to eliminate It.* Healthcare IT News. https://www.healthcareitnews.com/news/how-ai-bias-happens-and-how-eliminate-it

6 AI myths debunked. (2019, November 5). Gartner. https://www.gartner.com/smarterwithgartner/5-ai-myths-debunked

60 notable machine learning statistics: 2023 market share & data analysis. (2023). Finances Online. https://financesonline.com/machine-learning-statistics/

Stobierski, T. (2021, January 19). *Data wrangling: What it is & why it's important.* Harvard Business School Online.

https://online.hbs.edu/blog/post/data-wrangling

Study.com. (2023, January). *Productive teaching tool or innovative cheating?* https://study.com/resources/perceptions-of-chatgpt-in-schools

Sundar, S. S., Schmit, C., Villasenor, J., & The Conversation. (2023, April 3). *Audits, "soft laws" and "automation bias": 3 experts break down what it could take to regulate AI and how hard it will be.* Fortune. https://fortune.com/2023/04/03/how-to-regulate-ai-challenges-three-experts/

Thorbecke, C. (2023, June 28). *OpenAI, maker of ChatGPT, hit with proposed class action lawsuit alleging it stole people's data.* CNN Business. https://edition.cnn.com/2023/06/28/tech/openai-chatgpt-microsoft-data-sued/index.html

University of San Diego. (2021, April 2). *Artificial intelligence in finance [15 examples].* https://onlinedegrees.sandiego.edu/artificial-intelligence-finance/

Velazquez, R. (2023, October 9). *30 examples of AI in finance.* Built In. https://builtin.com/artificial-intelligence/ai-finance-banking-applications-companies

What is applied AI? (n.d.). Cognizant. https://www.cognizant.com/us/en/glossary/applied-ai

What is generative AI? (n.d.). NVIDIA. https://www.nvidia.com/en-us/glossary/data-science/generative-ai/

What is machine learning? (n.d.). Oracle. https://www.oracle.com/hk/artificial-intelligence/machine-learning/what-is-machine-learning/

Wheeler, T. (2023, June 15). *The three challenges of AI regulation.* Brookings. https://www.brookings.edu/articles/the-three-challenges-of-ai-regulation/

Zakaryan, V. (2022, April 28). *What are common misconceptions about AI?* Postindustria. https://postindustria.com/what-are-common-misconceptions-about-ai-machine-learning/